A Practical Guide to Maintenance Engineering

A Practical Guide to Maintenance Engineering

C. L. Dunlop BSc, MSc (Eng)
Consultant Maintenance Engineer

Butterworths
London   Boston   Singapore   Sydney   Toronto   Wellington

PART OF REED INTERNATIONAL P.L.C.

First published, 1990

© **Butterworth & Co (Publishers) Ltd, 1990**

Butterworth International Edition, 1990

ISBN 0 408 05284 8

British Library Cataloguing in Publication Data

Dunlop, C. L.
    Practical guide to maintenance engineering.
    1. Equipment. Maintenance & repair
    I. Title
    620.0046
ISBN 0-408-05272-4

Library of Congress Cataloging-in-Publication Data

Dunlop, C. L.
    A practical guide to maintenance engineering/
  C. L. Dunlop
      p.   cm.
    Includes bibliographical references.
    ISBN 0-408-05272-4
    1. Industrial equipment—Maintenance and repair.
    I. Title.
TS192.D86   1990
670.42′028′8—dc20
                                      90-1319
                                      CIP

Composition by Genesis Typesetting, Borough Green, Sevenoaks, Kent
Printed and (limp edition) bound in Great Britain by Page Bros. Ltd., Norwich, Norfolk. Cased edition bound by Hunter & Foulis Ltd., Edinburgh, Scotland

Contents

1 Equipment register

This system was created by an experienced hands on Engineer for Engineers or Managers designated to instigate such a system.

All too commom Planned and Preventive systems encountered by the undersigned were created by consultant academics with very little or no experience at all in the execution of such systems. Normally, their systems are so complex that they are doomed or they lack critical information to perfect the system. All too often, Consultants instigate a Planned and Preventive Maintenance system and hope that all discrepancies encuountered in its execution can be eliminated by skilled personnel assigned to that particular project.

If a Planned and Preventive Maintenance system is to be introduced, it must be simple but effective and should con-sist of the following three inter-related requirements.

(A)   A program of operation covering inspections adjustment, rectifications of faults and periodic overhauls.

(B)   A means of ensuring that these operations are carried out in accordance with the programmes.

(C)   A method of recording the work done and assessing the results.

As for any other planned activity, a Planned and Preventive Maintenance system must be controlled thereby necessitating some system of documentation. Where documentation is required to cater for maintenance work, a basically simple system will satisfy essential requirements providing the answers to four questions.

(A)   What is to be maintained?
(B)   How is it to be maintained?
(C)   When is it to be maintained?
(D)   Is it effective?

4

Taking these questions in order, the first step is the
compilation of the inventory to establish what has to be
maintained. This takes the form of an extensive list of
all PLANT and EQUIPMENT with identification codes, grouped
in locations and sub-locations within the recognised operating areas.

Each piece of plant and equipment coded will have an item
file which will positively identify it and record its history including cost of spares and manhours.

When the item to be maintained has been listed on the inventory codes, it should be established how the required maintenance is to be effected.

For this purpose, it is necessary to prepare a series of
master specifications separating the various periodical operations required by individual items.

It now remains to draw up the Maintenance Programme to establish when each item shall receive the specified attention
throughout the planning period. All maintenance items will
receive an occupation code as a guide to the item availability
on various documentations.

The maintenance schedules for the month give a history of items
overdue for inspection or overhaul at monthly intervals, while
the maintenance schedule historical presents a comprehensive
picture of maintenance requirements and their history for the
whole inventory showing all that has to be done to each item
and in which interval it must be accomplished.

However conscientiously the inventory specifications and maintenance programme is compiled and used, the scheme will be of
little value if it makes no provisions for the recording and
analysis of the results achieved.

This requires the use of feedback from the cards as issued,
maintenance schedules and feedback from the craftsman. All
history records will be analysed, acted on, and filed for
accurate information.

THE EQUIPMENT REGISTER

Introduction

I am constantly surprised while travelling throughout the world at how many organizations only have a general idea about their equipment. They may not even know the amount of equipment they have of a particular type, let alone the precise characteristics and requirements.

Maintenance Management is part of the maintenance science scenario and should be treated as such. It is critical in Maintenance Management that necessary information as applicable to their plant and equipment is available so that sound management policies are instigated; without this information Management is unlikely to instigate sound policies.

As plant investments escalate (Nelson's Index) and downtime creates unacceptable loss of production, the "demand style approach" in keeping plant available i.e. ˉFire Fightingˉ is recognised by perceptive management as another profit-draining area. Thus, a scientific management approach must be employed.

What is an Equipment Register?

An Equipment Register is a system which records all types of plant and equipment and its characteristics. Each piece of equipment will have a unique identification code which will be employed in the following areas:-

(1) A. Identification of company's assets.

     B. Location and function of all critical plant and equipment (i.e. no standby).

(2) Location and function of all independent plant and equipment, commissioned or non-commissioned.

(3) Identification of all system standby plant in the event of system failure (installed).

(4) Identification of all standby plant not installed i.e. Pump and motor mounted skid.

(5) Identification and location of redundant plant and equipment i.e. plant no longer utilized by production or maintenance.

(6)   The establishment of a planned maintenance system
        common to all operating companies which will include
        the following:-

        (A) Annual overhaul schedules
        (B) Preventive maintenance inspection schedules
        (C) Lubrication schedules

(7)   The establishment of spare part requirements
        "minimum and maximum stock holding" as applicable
        to all independent plant and equipment.

(8)   "Job Planning" - engineering standards and material
        standards.

(9)   Written on the maintenance work order to communicate
        to the repair man which equipment he is to work on.

(10)   Used to establish a maintenance service schedule
        for the equipment.

(11)   Serves as an account number to collect costs incurred
        by the equipment.

(12)   Identification of a data base record wherein informa-
        tion about equipment performance and costs are recorded.

(13)   Identification of equipment in management reports.

(14)   Identification of equipment in technical and cost
        studies.

(15)   Inter-company standardization i.e. plant and equipment
        management and staffing levels.

(16)   Identification and location of common spare parts
        and equipment applicable to all companies within
        the organization.

(17)   Spare stock Capital Cost Factor.

(18)   Common statutory inspection requirements.

(19)   Inter-company comparisons i.e. performance and produc-
        tion costs.

(20)   Identification of training requirements and the establishment of training facilities.

(21)   Establishment of necessary safety programs, and the instigation of Statutory Inspection schedules i.e. pressure vessels and safety equipment and plant.

(22)   New project plant and equipment standardization.

The operating company or companies will use common policies and procedures for equipment identification and will consist of the following codes:-

(A)   F.I.N. Code = Functional Identification Number. Identifies the process function and location.

(B)   S.P.I.N. Code = Specific Process Identification Number. Identifies the specific equipment item performing a particular process.

We believe the following elements must be identified in the F.I.N. Code.

(1)   Company or companies.

(2)   Major location (field, operating area, plant).

(3)   Minor location (cost centres within the major locations).

(4)   System fuel gas, cooling water, crude collection.

(5)   Equipment type (diesel engine, gas turbine).

(6)   Process function (sequential tag number).

(7)   Redundancy A.B.C.

Example of Gas turbine in Gas gathering located at the power plant in Location (A) Company (A)

Company	=	Company (A)	=	01
Major Location	=	Location (A)	=	03
Minor Location	=	Power Plant	=	01
System A	=	Gas gathering	=	G.G.
Equipment Type	=	Gas turbine	=	T.G.

Sequential = 5 digits
Standby Factor Identifier = A.B.C.

Resulting F.I.N. Number   = 01.03.01.G.G.T.G.(A)

We recommend the following structure for S.P.I.N. Code:-

(A)   Equipment Type - same as designated in the F.I.N. Code

(B)   Sequential No. - 00001.99999

EQUIPMENT CODE

DESCRIPTION	CODE
Absorber	AB
Anion Exchanger	AE
Air Filter	AF
Agitator	AG
Analyzer	AN
Bridge Crane	BC
Blow Down Drum	BD
Building	BL
Burner	BN
Boiler	BO
Blower	BW
Concentrator	CA
Cyclone	CC
Cation Exchanger	CE
Clarifier	CF
Centrifuge	CG
Chlorinator	CH
Chemical Injection System	CI
Condenser	CN
Cooler	CO
Cooling Tower	CT
Converter	CV
Conveyor	CY

DESCRIPTION	CODE
Oil Accumulator	OA
Oil Cooler	OC
Oil Filter	OF
Oil Reservoir	OR
Oil Separator	OS
Pump Centrifugal	PC
Purger	PE
Piping	PG
Pump Jack	PJ
Pump Reciprocating	PR
Pump Screw	PS
Pot	PT
Pump Vacuum	PU
Pump Vane	PV
Pump Well	PW
Pump Rotary	PY
Reboiler	RB
Reformer	RF
Regenerator	RG
Roller Track	RK
Reverse Osmosis Unit	RO
Package Air Conditioner	RP
Rectifier	RT

DESCRIPTION	CODE
Saturator	SA
Scrubber	SB
Scale	SC
Suction Drum	SD
Switch Gear	SG
Superheater	SH
Stack	SK
Silencer	SL
Separator	SP
Stripper	ST
Sump	SU
Turbine Gas	TG
Hydraulic Turbine	TH
Tank	TK
Transfer Pipe Line	TL
Trap	TP
Turbine Steam	TS
Tractor	TT
Tower	TW
Valve Block	VB
Valve Drain	VD
Vacuum Generator	VG
Vessel	VL
Valve Motor Operated	VM
Valve Pipeline	VP
Valve Relief	VR
Valve Safety	VS
Winch	WW

DESCRIPTION	CODE
Discharge Drum	DD
Desuperheater	DE
Dehumidifier	DF
Diverter Gate Valve	DG
Dehydration Unit	DH
Distillation Unit	DI
Dust Collector	DL
Demineralizer	DM
Drum	DR
Desulfurization	DS
Distributor	DT
Drip Well/Drip Sump	DW
Dryer	DY
Engine Diesel	ED
Engine Gas	EG
Exchanger Heat	EH
Ejector	EJ
Elevator	EL
Fluidizing Cooler	FC
Fluidizing Dryer	FD
Fork Lift	FV
Fluid Coupling	FK
Filter	FL
Fan	FN
Front Loader	FR
Flash Drum	FS

DESCRIPTION	CODE
Gear Box	GB
Gun Barrel	GL
Generator	GN
Heater	HE
Hydraulic Lift	HL
Hoist	HT
Inert Gas Generator	IG
Local Inst Pan Board	IL
Compressor Centrifugal	KC
Compressor Reciprocating	KR
Compressor Screw	KS
Compressor Rotary	KY
Lubricator	LB
Loader	LR
Motor Air	MA
Mixed Bed Exchanger	MB
Motor Electric	ME
Manifold	MF
Motor Starter	MS
Methnator	MT
Muffler	MV
Mixer	MX

14

Examples for illustration only.

(1)   Company Name

      00.   All companies combined
      01.   Company (A)
      02.   Company (B)
      03.   Company (C)

(2)   Major Location

      03.00   Location (A)
         01   Location (B)
         02   Location (C)
         03   Location (D)

(3)   Minor Location

      03.03.00   All Brega Terminal
            01   Power Plant
            02   Maintenance Workshop
            03   Docks and related facilities

(4)   System A

      G.C.   Crude gathering
      G.G.   Gas gathering
      G.O.   Gas-oil separation
      F.G.   Fuel gas

(5)   Equipment Type

      E.D.   Diesel engine
      T.G.   Gas turbine
      T.S.   Steam turbine
      K.C.   Centrifugal compressor

VE	Vacuum Ejector
VP	Valve Positioner
XC	Waugh Control
CD	Fire Detector
XE	Extinguisher System
ZB	Zener Barrier (Safety)

16

LOCATION CODES

01	Crude Metering, Reducing Area, Tank Farm
02	Water System, Chlorinators, Fire Alarm, Power House
08	Bunker Area
52	Government Gas Azaiba/MOD
54	Government Gas Sohar

INSTRUMENT FUNCTION TERMINOLOGY

AC	Control Component
AD	Display Component
AI	Input Component
AO	Output Component
AP	Process Component
AR	Power Supply
AS	Input/Output Component
AT	Test Equipment
CB	Batch Counter
CP	Prover Counter
CV	Control Valve
FH	Flow High Frequency Pick-up
FL	Flow Low Frequency Pick-up
FP	Flow Primary
FQ	Flow Counter
FS	Flow Switch/Indicator
FT	Flow Transmitter
ID	Interface Detector
IL	Local Indicator
IM	Instrument Motor
IP	Current to Pressure Convertor

LC	Level Controller
LP	Level Primary Element
LS	Level Switch
LT	Level Transmitter
MS	Miniature Switch Assembly
PA	Pulse Amplifier
PC	Pressure Controller
PD	Differential Pressure Switch/Indicator
PP	Primary Element
PR	Pressure Regulator
PS	Primary Sensor/Pressure Switch
PT	Pressure Transmitter
PV	Poppet Valve
QT	Analyser/Densitometer
SC	Scanner Assembly
SV	Solenoid Valve
TP	Temperature Primary Element
TS	Temperature Switch
TT	Temperature Transmitter
UD	Ball Detector
US	Limit Switch

NUMBERING SYSTEM AS APPLICABLE TO INSTRUMENTATION

The instrument numbering format is as follows:-

NNN/NN/AN---/NN/AA/N

where:

NNN	is	– system code (Numeric)
NN	is	– location code (Numeric)
AN---	is	– loop function (Alpha-Numeric)
NN	is	– loop serial number (Numeric)
AA/N	is	– instrument function and serial number (Alpha-Numeric)

eg. 117-02-PR005-01-TT-1

Loop Function

The loop function number identifies the main function of the loop along with its sequential assigned number. It is an Alpha-Numeric number in a 5 spaces format.
eg. PR005 is a pressure measurement loop number.

Loop Serial Number

The loop serial number identifies the sequential number of the loop in a particular location.
eg. PR005-01 is the first pressure measurement loop in a given location.

Instrument Function/Serial Number

The instrument function-serial number identifies the function of the instrument and its sequential number.
eg. TT-1 is the temperature transmitter number one.

Usage

This numbering system is designed to facilitate the identification and registration of instruments in EMC department, and for Maintenance Planning purposes.

LOOP FUNCTION TERMINOLOGY

AL	Alarm Loop
AR	Central Power Supply Loop
FR/FRXA/FM	Flow Measurement Loop
H	Heater Control Loop
IL	Local Indicator
K	Compressor Control Loop
LT/LRXA/LA/LC	Level Measurement Loop
MV	Motorised Valve Loop
P	Pump Control Loop
PR/PC/PA	Pressure Measurement Loop
QT	Density Measurement Loop
TR/TRC	Temperature Measurement Loop
XA	Fire Alarm Loop
QC	Chlorine Control

EQUIPMENT REGISTER SOURCE FORM

FORM ID

FORM SEQUENCE NO. ACT

P M E R 0 1

SPIN ————

EQUIPMENT FIN

CRIT GROUP SERIAL

A A

EQUIPMENT NOUN NAME

EQUIPMENT SHORT DESCRIPTION

B A

EQUIPMENT LONG DESCRIPTION

C A
C B
C C
C D
C E
C F
C G
C H
C I
C J
C K
C L
C M
C N
C O

EQUIPMENT LOCATION

D A
D B

MANUFACTURER'S EQUIPMENT SERIAL NO.

PACKAGE/UNIT SERIAL NO.

E A

PURCHASE ORDER NO.

P. O. DATE   LINE.

F A

MANUFACTURER'S CODE/NAME/ADDRESS

COMMODITY NO.   TAG NO.   ASSET NO.

CURR   UNIT PRICE   INSTALLATION DT   COMMISSION DT

VENDOR'S CODE/NAME/ADDRESS

G A
G B
G C
G D
G E
G F

EQUIPMENT REGISTER SOURCE FORM

FORM ID
P M E R 0 1

FORM SEQUENCE NO. ACT
EQUIPMENT FIN
_ _ _ _

CRIT GROUP SPIN SERIAL
V FC 000009

A A 0203 03GQPD07A

EQUIPMENT NOUN NAME

B A TURBINE GAS

EQUIPMENT SHORT DESCRIPTION

EQUIPMENT LONG DESCRIPTION

C A	TYPE MODEL FRAME 5	
C B	OUTPUT 13500 KW	
C C		TEMP OUT 895 DEGREES F
C D	TEMP IN 80 DEGREES F	PRESS OUT 14.17 PSIA
C E	PRESS IN 14.17 PSIA	NOS. OF SHEETS 1
C F		STAGES OF COMPRESSOR 16
C G	CYCLES 60 HZ	SPEED COMPRESSOR 5100 RPM
C H	STAGES TURBINE 2	
C I	SPEED TURBINE 5100 RPM	
C J		RATING 350 HP AT 1800 RPM
C K	CONTROL SYSTEM GE-DC	440 VOLTS 3 PHASE
C L	STARTING MOTOR TYPE K	
C M	MAXIMUM TEMP RAISE (STATOR/ROTOR)	
C N	60185 DEGREES C	
C O		

EQUIPMENT LOCATION

D A DAHRA CENTRAL AREA POWER STATION
D B OASIS OIL COMPANY

MANUFACTURER'S EQUIPMENT SERIAL NO.

E A 179119

MANUFACTURER'S EQUIPMENT SERIAL NO.   PACKAGE/UNIT SERIAL NO.   COMMODITY NO.   TAG NO.   ASSET NO.

PURCHASE ORDER NO.   P. O. DATE   LINE   CURR   UNIT PRICE   INSTALLATION DT   COMMISSION DT

F A

MANUFACTURER'S CODE/NAME/ADDRESS   VENDOR'S CODE/NAME/ADDRESS

G A	GE
G B	
G C	
G D	
G E	
G F	

| EQUIPMENT NO | EQUIPMENT RECORD CARD | DRAWING NO |
| LOCATION CODE | | MANUAL NO |

EQUIPMENT NAME

| MANUFACTURER | ADDRESS | TELEX NO |
| | | PHONE NO |

| STATUTORY INSPECTION CLASSIFICATION A B C | PRIORITY CODE A B C |

PLANT DESCRIPTION

SERVICE DESCRIPTION

| INSURANCE COMPANY | TELEX NO | COMMISSIONED DATE |
| | PHONE NO | |

| SAFETY CODE NO FREQUENCY | LUBRICATION CODE NO FREQUENCY |

| MAINTENANCE PLANNED JOB NO | MAINTENANCE PREVENTIVE JOB NO |

| COST CENTRE CAPITAL NO REVENUE NO | STANDBY PLANT AVAILABLE YES NO |

REMARKS

DATE PREPARED

24

Left card:

EQUIPMENT NO.   503 0411 KGT 301   TURBINE, GAS   INJECTION COMPRESSOR GAS TURBINE - 27950 HP

IDENTIFIERS
Assembly/Instr. No.   02 30 11 224 1
Commodity Code   244382
Mfg. Serial No.   M788864651
Co. Asset No.   KGT-301
Bechtel No.

MANUFACTURER
GENERAL ELECTRIC COMPANY
GAS TURBINE PRODUCTS DIV.
SCHENECTADY, NY 12345

VENDOR

LONG DESCRIPTION
TURBINE, GAS
HP (AT 10 FEET ALT): 27950   TEMP (OUT): 956 DEG-F
TEMP (IN): 90 DEG-F   PRESSURE (OUT): 14.81 PSIA
PRESSURE (IN): 14.56 PSIA   SHAFTS: 2
CYCLE: SIMPLE   CASING SPLIT: HORIZONTAL
ROTATION: CCW   RPM (COMPRESSOR): 5100
RPM (TURBINE): 4670   STAGES (COMPRESSOR): 16
STAGES (TURBINE): 2
CONTROL SYSTEM: SPEEDTRONIC
STARTING MOTOR: GE CUSTOM 8000 ELECTRIC
GE MODEL/SERIES: MS-5002-B

DRIVES: INJECTION COMPRESSOR - DRESSER/CLARK - 272B4/4
DRIVEN EQUIP NO: K-301

P.O. INFORMATION
Number   62.10-KGT-3
Date   09/17/76
Line Number   1
Unit Cost $000000.00

CRITICALITY   1
LOCATION   CLUSTER III - ADJACENT TO INJECTION COMPRESSOR

NOTES:

COMPUTER PRINTOUT AS APPLICABLE TO FUNCTIONAL IDENTIFICATION NUMBER (F.I.N. CODE)

Right card:

EQUIPMENT NO.   503 0411 KGT 301   EXCHANGER, HEAT   GAS TURBINE LUBE OIL HEAT EXCHANGER

IDENTIFIERS
Assembly/Instr. No.   KGT 301 E 1
Commodity Code   16 05 44 367 2
Mfg. Serial No.   6-62238-01-4
Co. Asset No.   M788864651
Bechtel No.   KG-301-E1

MANUFACTURER
AMERICAN STANDARD CORP.
HEAT TRANSFER DIV.
POWER & CONTROLS GROUP
BUFFALO, NY 12467

VENDOR

LONG DESCRIPTION
EXCHANGER, HEAT
MAX WP (SHELL): 150 PSI   MAX WP (TUBE): 150 PSI
MAX TEMP (SHELL): 250 DEG-C   MAX TEMP (TUBE): 250 DEG-C
CAPACITY: 577 GPM
MEDIUM COOLED: OIL
AM STD TYPE/MODEL: SHELL & TUBE/C-3000

P.O. INFORMATION
Number   62.10-KGT-3
Date   09/17/76
Line Number   37
Unit Cost $000000.00

CRITICALITY   2
LOCATION   CLUSTER III - GAS TURBINE KGT-301

NOTES:

25

KEY	ITEM	EQUIPMENT NO	ASSEMBLY NO	NAME	DESCRIPTION
SYSTEM: 5930820	1200	593 0820 PBA 301		CHARGER, BATTERY	BATTERY CHARGER – SUB STA 3-1
	1201	593 0820 PMC 301 A		CENTER, MOTOR CTL	5 KV CLASS, MOTOR CONTROL CENTER
	1202	593 0820 PMC 301 B		CENTER, MOTOR CTL	5 KV CLASS, MOTOR CONTROL CENTER
	1203	593 0820 PMC 302 A		CENTER, MOTOR CTL	MOTOR CONTROL CENTER, 400 V
	1204	593 0820 PMC 302 B		CENTER, MOTOR CTL	MOTOR CONTROL CENTER, 400 V
	1205	593 0820 PMC 303 A		CENTER, MOTOR CTL	MOTOR CONTROL CENTER, 400 V
	1206	593 0820 PMC 303 B		CENTER, MOTOR CTL	MOTOR CONTROL CENTER, 400 V
	1207	593 0820 PMC 304 A		CENTER, MOTOR CTL	MOTOR CONTROL CENTER, 400 V
	1208	593 0820 PMC 304 B		CENTER, MOTOR CTL	MOTOR CONTROL CENTER, 400 V
	1209	593 0820 PMC 305 A		CENTER, MOTOR CTL	MOTOR CONTROL CENTER, 400 V
	1210	593 0820 PMC 305 B		CENTER, MOTOR CTL	MOTOR CONTROL CENTER, 400 V
	1211	593 0820 PMC 306		CENTER, MOTOR CTL	MOTOR CONTROL CENTER, 400 V
	1212	593 0820 PSG 301		SWITCHGEAR	5 KV CLASS SWITCHGEAR
	1213	593 0820 PSG 302		SWITCHGEAR	SWITCHGEAR, 400 V
	1214	593 0820 PSG 303		SWITCHGEAR	SWITCHGEAR, 400 V
	1215	593 0820 PSG 304		SWITCHGEAR	SWITCHGEAR, 400 V
	1216	593 0820 PSG 305		SWITCHGEAR	SWITCHGEAR, 400 V
	1217	593 0820 PXF 302 A		TRANSFORMER, POWER	TRANSFORMER, 3.15 KV – 400 Y/230 V
	1218	593 0820 PXF 302 B		TRANSFORMER, POWER	TRANSFORMER, 3.15 KV – 400 Y/230 V
	1219	593 0820 PXF 303 A		TRANSFORMER, POWER	TRANSFORMER, 3.15 KV – 400 Y/230 V
	1220	593 0820 PXF 303 B		TRANSFORMER, POWER	TRANSFORMER, 3.15 KV – 400 Y/230 V
	1221	593 0820 PXF 304 A		TRANSFORMER, POWER	TRANSFORMER, 3.15 KV – 400 Y/230 V
	1222	593 0820 PXF 304 B		TRANSFORMER, POWER	TRANSFORMER, 3.15 KV – 400 Y/230 V
	1223	593 0820 PXF 305 A		TRANSFORMER, POWER	TRANSFORMER, 3.15 KV – 400 Y/230 V
	1224	593 0820 PXF 305 B		TRANSFORMER, POWER	TRANSFORMER, 3.15 KV – 400 Y/230 V
	1225	593 0820 PXF 306		TRANSFORMER, POWER	TRANSFORMER, 3.15 KV – 400 Y/230 V
	1226	593 0820 PXF 307		TRANSFORMER, POWER	TRANSFORMER, 3.15 KV – 600 V
	1227	593 0820 R 302		CONDITIONER, AIR	CARRIER AIR CONDITIONER-SUB STA 3-1
SYSTEM: 5930910	1228	593 0911 B 302		TANK	FUEL GAS SYSTEM HEATER
	1229	593 0911 B 352		TANK	FUEL GAS SYSTEM HEATER
	1230	593 0911 D 309		DRUM	FUEL GAS SYSTEM KNOCK-OUT DRUM-TR. 1
	1231	593 0911 D 310		SEPARATOR	FUEL GAS SEPARATOR – TR. 1
	1232	593 0911 D 359		DRUM	FUEL GAS SYSTEM KNOCK-OUT DRUM-TR. 2
	1233	593 0911 D 360		VESSEL	FUEL GAS SEPARATOR – TR. 2
SYSTEM: 5930920	1234	593 0921 K 303 A		COMPRESSOR, RECIP.	INST. & UTILITY AIR COMPRESSOR
	1235	593 0921 K 303 B		COMPRESSOR, RECIP.	INST. & UTILITY AIR COMPRESSOR
	1236	593 0921 K 303 C		COMPRESSOR, RECIP.	INST. & UTILITY AIR COMPRESSOR
	1237	593 0921 KM 303 A		MOTOR, ELECTRIC	INST AND UTILITY AIR COMP MOTOR-100 HP
	1238	593 0921 KM 303 B		MOTOR, ELECTRIC	INST AND UTILITY AIR COMP MOTOR-100 HP
	1239	593 0921 KM 303 C		MOTOR, ELECTRIC	INST AND UTILITY AIR COMP MOTOR-100 HP
	1240	593 0922 E 309 A		EXCHANGER, HEAT	COMPRESSED AIR INTERCOOLER
	1241	593 0922 E 309 B		EXCHANGER, HEAT	COMPRESSED AIR INTERCOOLER
	1242	593 0922 E 309 C		EXCHANGER, HEAT	COMPRESSED AIR INTERCOOLER
	1243	593 0922 E 310 A		EXCHANGER, HEAT	COMPRESSED AIR AFTERCOOLER
	1244	593 0922 E 310 B		EXCHANGER, HEAT	COMPRESSED AIR AFTERCOOLER
	1245	593 0922 E 311 A		EXCHANGER, HEAT	AIR COMPRESSOR CW AIR COOLER
	1246	593 0922 E 311 B		EXCHANGER, HEAT	AIR COMPRESSOR CW AIR COOLER
	1247	593 0922 EM 311 A		MOTOR, ELECTRIC	AIR COMP. CW COOLER MOTOR – 15 HP
	1248	593 0922 EM 311 B		MOTOR, ELECTRIC	AIR COMP. CW COOLER MOTOR – 15 HP
	1249	593 0922 Y 301 A		DRYER, AIR	INSTRUMENT AIR DRYER
	1250	593 0922 Y 301 B		DRYER, AIR	INSTRUMENT AIR DRYER
	1251	593 0922 Y 302 A		DRYER, AIR	INSTRUMENT AIR DRYER

2 Planned preventive and running maintenance

PLANNED PREVENTIVE AND RUNNING MAINTENANCE

OBJECTIVES AND GUIDELINES

1.1 For minimum cost, maintenance must be managed in the broadest maintenance sense. The basic objectives of maintenance are:-

    (a) To manage the Maintenance Department so as to obtain minimum total operating costs.

    (b) To keep facilities and equipment operating in good condition.

    (c) To keep facilities and equipment operating the optimum percentage of the time.

The maintenance division alone without help from other departments cannot achieve these objectives. Management must give attention not only to the control of factors internal to maintenance, but also to the control of external factors such as operations, logistics, engineering, and inspection.

For example, if operations insists that a piece of equipment operates the maximum percentage of time, the maintenance division cannot keep it operating the optimum percentage of time (lack of maintenance period).

1.2 Some of the basic principles of optimum maintenance are:-

1.3 Maintenance should be considered an integral part of the organization handling one phase of operations and as important as operations. Basically, it is a service function and, as such, should not be allowed to dominate operations.

1.4 Maintenance must be managed in a manner commensurate with its importance. Management should devote as much time to it as to operations.

1.5   Maintenance work must be controlled at its source. A single individual, usually a maintenance foreman, must be responsible for a facility and must control its maintenance cost. Only this individual or a supervisor in the direct chair of command can authorize work on the facility.

1.6   The work load must be controlled so that work is balanced in relation to manpower, manpower is utilized in the most efficient manner, and personnel are kept at minimum levels. Backlog must be determined periodically to accomplish this.

1.7   There must be an orderly execution of the work load. Most often this is accomplished with a work order control system utilizing written work orders. Such systems are designed around the seven basic functions.

      (a)   Request
      (b)   Plan
      (e)   Estimate
      (d)   Authorize
      (e)   Schedule
      (f)   Execute
      (g)   Review

1.8   Except for unusual circumstances such as emergencies, maintenance work is planned before work starts by someone other than the foreman - Maintenance Engineer and Maintenance Supervisor, Running Maintenance Supervisor.

1.9   The originating Supervisor or Engineer obtains a cost estimate before authorizing routine maintenance work.

2.0   The work of every maintenance man is scheduled and conditions are such that these schedules are met a high percentage of the time.

2.1   The maintenance organization structure is designed around three levels of maintenance:-

       (a)   <u>First level</u> - station keeping (Operations) routine on-site work, such as periodic lubrication cleaning, checking, and small adjustments.

       (b)   <u>Second level</u> - field or running maintenance crews or crew are seconded to the equipment for preventive or corrective maintenance.

       (c)   <u>Third level</u> - planned and preventive maintenance, planned non-availability, overhaul, repair fabrication, assembly.

2.2   Foremen have three basic responsibilities:-

       (a)   Obtain high quality work.

       (b)   Obtain satisfactory labour productivity.

       (c)   Minimize lag time, regarding spaces stock situation. Provide feed back man hour, spares used.

2.3   All maintenance functions are reviewed automatically preferably by exception, and compared to standard function. Action is instigated to correct variances.

2.4   Overall maintenance departments performance is compared routinely to established indexes. Action is instigated to correct variances.

2.5   Maintenance costs are reported in such a way that the cost of maintaining, and non-availability factor on major equipment can be determined easily (work order).

2.6   Maintenance receives adequate technical support. Drawings, shop plans, and machine specifications, are readily available.

2.7   Maintenance has a program of measuring labour productivity, analyzing performances, performing methods studies, preparing standards, and performing other staff technical functions.

2.8   Maintenance has a program for monitoring equipment conditions, preferably on an exception basis. Variances in an equipment's failure frequency, spare parts usage, performance, or maintenance costs, are investigated and corrective action taken.

2.9   There is a centralized preventive maintenance program. It is optimized in relation to equipment reliability and the most economic service factor. Enough but not too much preventive maintenance is performed.

3.0   There is flexible policy for the use of contractors to cover peak work requirements. The maintenance department is not staffed to perform all work.

3.1   Policy permits and procedures exist for using outside technical experts and consultants to solve specific maintenance problems.

3.2   Maintenance has a craft training program which both teaches new skills and upgrades existing skills.

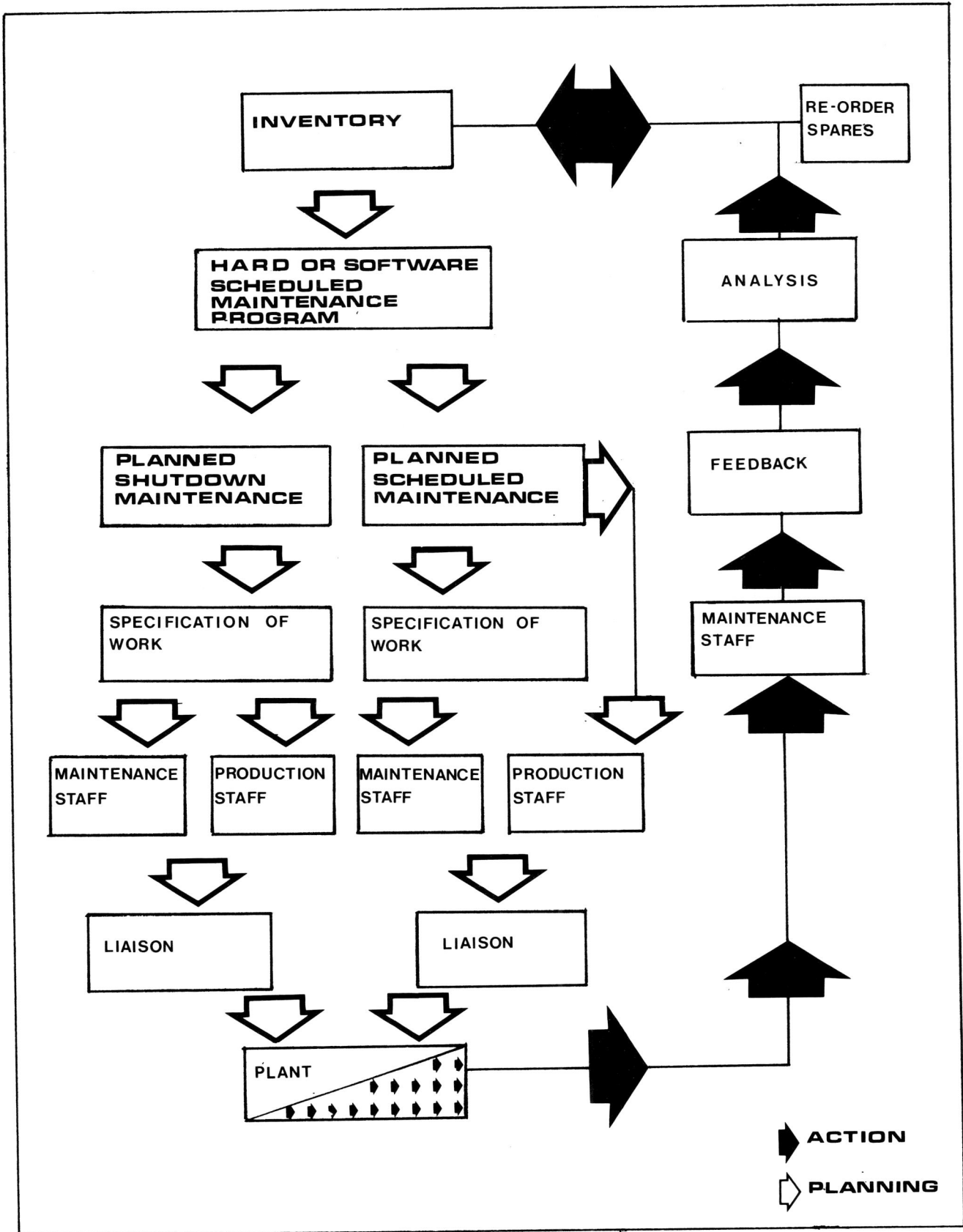

```
INVENTORY ──────────◄►────────── RE-ORDER
                                  SPARES
    │                                ▲
    ▼                                │
HARD OR SOFTWARE                 ANALYSIS
SCHEDULED
MAINTENANCE                          ▲
PROGRAM                              │
    │           │                 FEEDBACK
    ▼           ▼                    ▲
PLANNED      PLANNED                 │
SHUTDOWN     SCHEDULED ──►        MAINTENANCE
MAINTENANCE  MAINTENANCE          STAFF
    │           │        │           ▲
    ▼           ▼        ▼           │
SPECIFICATION  SPECIFICATION
OF WORK        OF WORK

MAINTENANCE  PRODUCTION  MAINTENANCE  PRODUCTION
STAFF        STAFF       STAFF        STAFF

    ▼                        ▼
  LIAISON                  LIAISON

         ▼        ▼
         PLANT ──────────►
```

ACTION

PLANNING

34

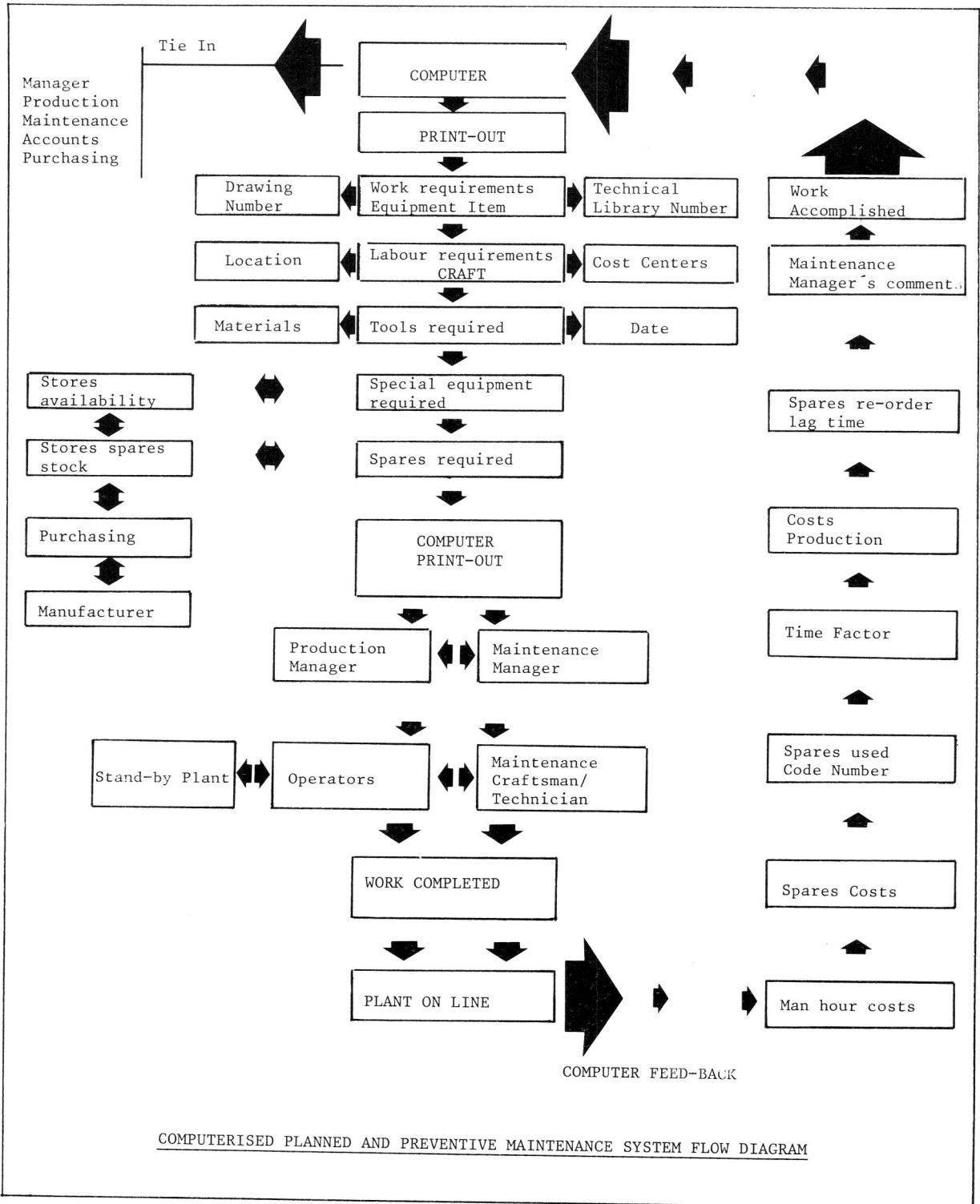

COMPUTERISED PLANNED AND PREVENTIVE MAINTENANCE SYSTEM FLOW DIAGRAM

EQUIPMENT NO 030001GTTG LOCATION CODE Area 41.C(2)	EQUIPMENT RECORD CARD	DRAWING NO COB A/21 MANUAL NO COB.6056

EQUIPMENT NAME

Cooper Rolls Gas Turbine Model 6056

MANUFACTURER Cooper Rolls	ADDRESS Mount Vermon Ohio 43050, U.S.A.	TELEX NO 24.5388 PHONE NO 614.397
STATUTORY INSPECTION CLASSIFICATION A B C Code B	PRIORITY CODE A B C Priority A	

PLANT DESCRIPTION   Gas Turbine

Power = B.H.P. 30550
      K.W.   22780

Speed =   4950 R.P.M.
Fuel Rate =   B.T.U./HPH 7312
             KJ/KWH 10343
Thermal Efficiency % = 34.8
Overall Weight lbs. = 53000
Rolls Royce Gas Generator = R.B.211
G.G. Speed = 6664(4)
G.G. Compression Ratio = 18.4:1
G.G. Compressor Stages = 7.6
G.G. Turbine Stages = 1.1

G.G. Exhaust Temp. °F = 835
G.G. Exhaust Mass flow °C = 446
                            = 207 lbs. sq."

G.G. Combustors = 1^2
G.G. Turbine Inlet Temp. = 2015
Cooper Bessmer Model RT56
Power Turbine Stages 2

SERVICE DESCRIPTION

Gas re-injection C Field

INSURANCE COMPANY Lloyds London E12 Cheapside, U.K.	TELEX NO 43271 PHONE NO 01.9624898	COMMISSIONED DATE November 21, 1981
SAFETY CODE NO A1 FREQUENCY	LUBRICATION CODE NO 44A FREQUENCY 4=P	
MAINTENANCE PLANNED JOB NO A23246L	MAINTENANCE PREVENTIVE JOB NO 432S/44	
COST CENTRE CAPITAL NO 2323 REVENUE NO 44?21	STANDBY PLANT AVAILABLE YES X NO. 04001GTT6	

REMARKS

Prepare and place Standby Unit No. 04001GTT6 in service before attempting any servicing.

DATE PREPARED

MAINTENANCE

The aim of a Maintenance Department should be to provide an efficient service in order to achieve as high a plant availability as possible at the cheapest cost.

To achieve the above, periodic servicing must take place and normally falls under the following items:-

(1)   Planned Maintenance - Major repairs, overhaul calibration (planned factor).

(2)   Preventive Maintenance - Necessary servicing of plant or equipment to prevent failure (corrective factor).

(3)   Emergency Maintenance - Repair or rectification as soon as possible depending on failure.

Clearly, if high plant availability is to be achieved, as much maintenance work should be accomplished while the plant is in operation. Units should only be shut down to rectify a particular defect if that defect affects the safety of personnel or plant or leads inevitably to loss of production. First aid or emergency maintenance measures should be put into operation whenever possible to keep the plant running. Defects requiring a unit outage but not causing embarrassment should be left until a sufficient number have occurred and the plant then taken off load at a suitable production time.

Plant efficiency is also a major object of the Maintenance Department and it is the duty of the Maintenance Manager to keep a close liaison with the Planning and Operational Departments in observing trends in plant performance. On the basis of these observations, maintenance should be arranged on plant falling short of its rated efficiency so that the cost of operation and maintenance is minimised.

A close watch must also be kept for faults which are inherent in the design of any equipment. These are generally known as type faults. They should be carefully studied - (Equipment Failure Sheet - 6M4) so that modifications may be introduced until satisfactory solutions are reached.

PLANNED MAINTENANCE SERVICE SHEET (1MF)

(1.1)   The Planned Maintenance Service Sheet (1MF) is invaluable
        for accomplishing control of any maintenance function.
        This sheet is issued in conjunction with the following
        documentations:-

          Planned Maintenance Task Sheet (2MD1)
          Preventive Maintenance Task Sheet 5M(2)
          Maintenance Trouble Shooting Guide (3M)
          Statutory Inspection Task Sheet (61N)
          Lubrication Schedule
          Operation or Production Defect Card (211)

(1.2)   The Service Sheet is issued by the respective Supervisory
        Maintenance personnel with the following items completed:-

          Equipment Name
          Job Number
          Job Function and Job Location
          Work Description
          Craft
          Work Permit information
          Safety Instructions

(1.3)   The document is then passed on to the Foreman or Supervisor
        who completes the following sections:-

          Tools required
          Materials required
          Special Tools required
          Spares required
          Special instructions

(1.4)   The Craftsman or Technician receives the part completed
         Planned Maintenance Service Sheet with the relevant
         document (Preventive Maintenance Task Sheet 5M2) and
         completes work as designated. On completion of the work
         the craftsman or technician completes the following:-

                    Spares used - code number
                    Manhours by Craft
                    Total Craft hours
                    Any remarks as necessary
                    Work completion date

(1.5)   The completed Service Sheet is then returned to the
         Maintenance Supervisor or Foreman who enters the following
         information:-

                    Cost Center
                    Planned or Un-planned maintenance
                    Date scheduled
                    Date completed
                    Lubrication Code (if any)
                    Statutory Inspection Code (if any)
                    Spares cost
                    Man-hour cost
                    Machine availability factor
                    Supervisor's remarks (if any)
                    Date & Signature

(1.6)   A copy of the completed form is submitted to the following
         departments for necessary action:-

                    Accounts
                    Stores (min - max stock holding)
                    Purchasing

PLANNED MAINTENANCE SERVICE SHEET	1 M F

EQUIPMENT

DATE PREPARED	(JOB NO)

JOB FUNCTION | **JOB LOCATION**

WORK DESCRIPTION	CRAFT	☐ M ☐ E ☐ I ☐ O

SAFETY INSTRUCTIONS	WORK PERMIT	PERMIT INFORMATION
	YES ☐	NO
	NO ☐	ISSUED BY
	DURATION HRS	DATE

TOOLS REQUIRED	MATERIALS	SPECIAL TOOLS
1 2 3 4 5	1 2 3 4 5	1 2 3 4 5

SPECIAL INSTRUCTIONS	SPARES REQUIRED
	1 2 3 4 5

TECHNICIAN CRAFTSMAN REMARKS	MANHOURS M E I O	SPARES USED 1 2 3 4 5

WORK COMPLETION	■ TOTAL HOURS

SUPERVISORS   REMARKS

COST CENTER	
PLANNED ☐ YES ☐ NO	
UN - PLANNED ☐ YES ☐ NO	
DATE SCHEDULED	
DATE COMPLETED	
LUBRICATION CODE NO	
STATUTORY INSPECTION NO	
SPARES COST	
MANHOUR COST	
AVAILABILITY FACTOR	

SUPERVISOR

DATE

40

FEEDBACK DOCUMENTATION

Suitable Feedback Documentation is essential for any Planned or Preventive Maintenance system in order to achieve the following:-

(A)   Summarised details of adjustments made, failures, and action taken to rectify such conditions. The cause of breakdowns where known and results to alleviate such conditions. The manhours devoted to the maintenance operation and the item downtime involved including costs.

        The history is the "Memory" of the planned maintenance scheme, and critical study of the data it provides will give the clearest measure of the efficiency of methods used.

(B)   The information to be recorded can be retrieved from the Planned Maintenance Service Sheet (1MF).

(C)   Demands for unscheduled maintenance should become progressively fewer as the scheme is perfected and the cause of failure is identified and eliminated. If they persist on a large scale, this will indicate areas in which the maintenance is inadequate. If, on the other hand, the downward trend continues to the point where there is virtually no repairs, this may indicate uneconomical "over maintenance".

(D)   The curve of maintenance costs elevates very rapidly as the incidence of failure approaches zero and the cost of an occasional breakdown may in fact be less than that of the maintenance necessary to prevail it. The purpose of this "History Records" is to show not only where maintenance is inadequate, but also where less will suffice.

PREVENTIVE MAINTENANCE TASK SHEET (5M2)

The Preventive Maintenance Task Sheet (5M2) is mainly utilised in the function of minor repairs, lubrication and calibration of plant between major overhauls.

It contains the basic information as applicable to the Planned Maintenance Task Sheet but is less comprehensive in the section of Tasks To Be Performed.

The Safety section is identical to the Planned Maintenance Task Sheet and as important.

The Preventive Maintenance Task Sheet is also issued in conjunction with Planned Maintenance Service Sheet (IMF).

42

PREVENTIVE MAINTENENCE TASK SHEET		5M(2)

EQUIPMENT NAME | JOB NO

EQUIPMENT NO | CRITICALITY | SHEET NO

CRAFT | FREQUENCY | MANUAL REFERENCE NO

SAFETY   WORK PERMIT ☐ YES ☐ NO

HOT WORK ☐   GOGGLES ☐   BREATHING EQUIP ☐   LIFE LINE ☐

PROTECTIVE SUIT ☐ GLOVES ☐ HEARING PROTECTION ☐ GAS TEST ☐

TASK NO	TASKS TO BE PERFORMED	REMARKS

SPECIAL INSTRUCTION

DATE PREPARED   MANUFACTURER   INSTALLED

EQUIPMENT FAILURE SHEET (NO. 6M4)

The Equipment Failure Sheet is a document which is normally located at the Plant or Equipment being monitored. Information acquired from a completed Equipment Failure Sheet is invaluable as it will determine reason or reasons for a low availability factor.

This document will supply the following basic information:-

(A)   Establish fault areas on plant or equipment to uncover conditions leading to production breakdowns or harmful depreciations.

(B)   Sterilize machine or plant conditions to adjust or repair conditions while conditions are at a minor stage.

(C)   To computerize machine or plant percentages availability factor.

(D)   To assist in the maximum and minimum spare stock factor.

(E)   Formation of Planned and Preventive Maintenance schedules to overcome such conditions.

(F)   Assist in the re-design of fault area as necessary.

It has been the writer's experience that the Equipment Failure Sheet is invaluable for the identification of breakdown peaks in the failure of plant and equipment.

It has also been demonstrated that plant availability can be improved over a short period by elevating the failure peaks.

EQUIPMENT FAILURE SHEET (6M4)

EQUIPMENT NAME _____

CRITICALITY _____ ☐ A ☐ B ☐ C

EQUIPMENT NO

LOCATION CODE

DATE	TIME OF FAILURE	DESCRIPTION	REPAIRS COMPLETED SPARES USED	SPARE NO	TIME ON	COMMENTS OF CRAFTSMAN	SUPERVISOR REMARKS

NON AVAILABILITY (TOTAL HOURS)   TOTAL SPARES COST   TOTAL MANHOUR COST

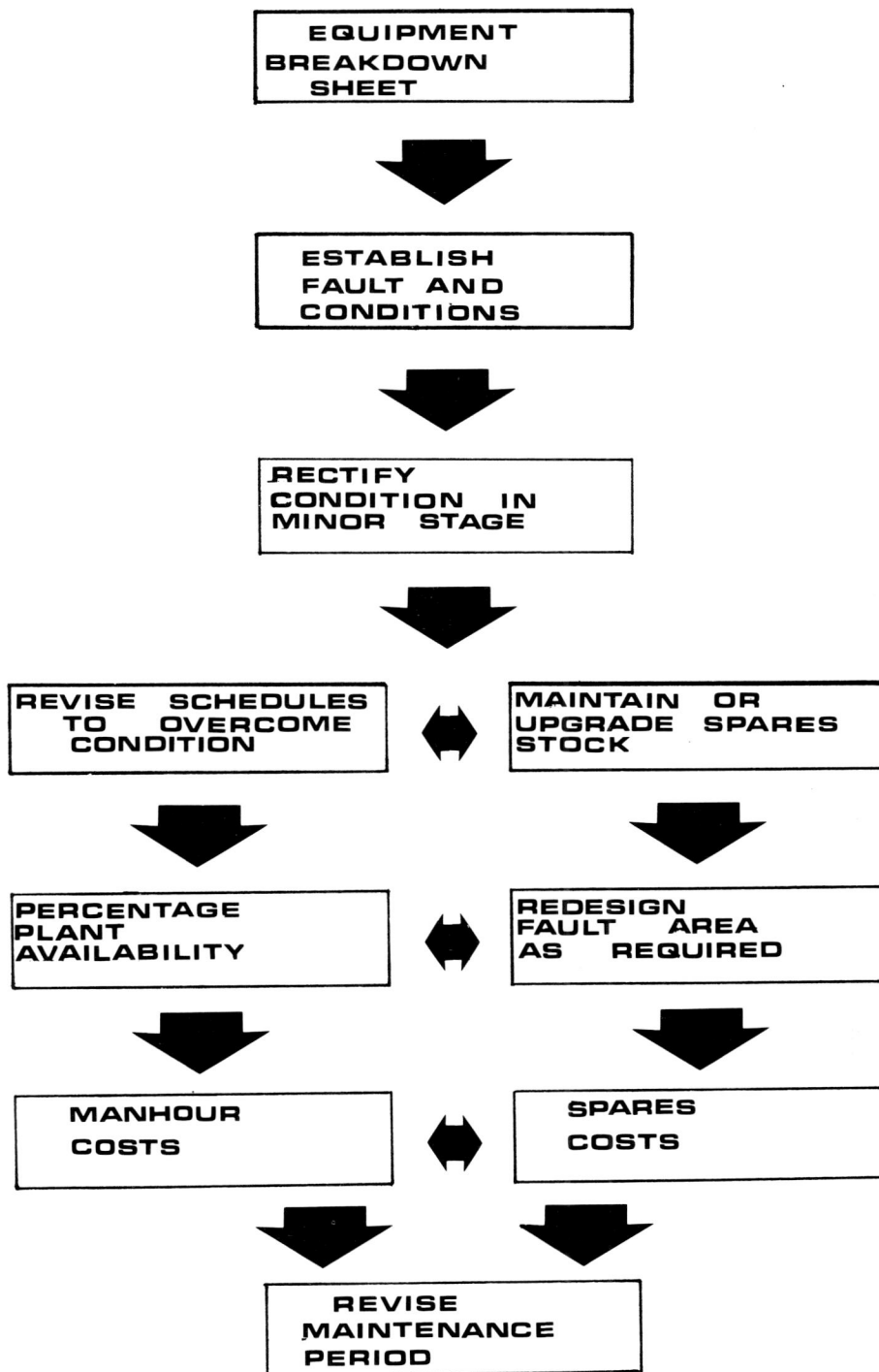

```
            ┌─────────────────┐
            │    EQUIPMENT    │
            │   BREAKDOWN     │
            │     SHEET       │
            └─────────────────┘
                     ▼
            ┌─────────────────┐
            │   ESTABLISH     │
            │  FAULT AND      │
            │  CONDITIONS     │
            └─────────────────┘
                     ▼
            ┌─────────────────┐
            │  RECTIFY        │
            │  CONDITION IN   │
            │  MINOR STAGE    │
            └─────────────────┘
                     ▼
```

REVISE SCHEDULES TO OVERCOME CONDITION	◄►	MAINTAIN OR UPGRADE SPARES STOCK
PERCENTAGE PLANT AVAILABILITY	◄►	REDESIGN FAULT AREA AS REQUIRED
MANHOUR COSTS	◄►	SPARES COSTS

```
            ┌─────────────────┐
            │    REVISE        │
            │  MAINTENANCE     │
            │    PERIOD        │
            └─────────────────┘
```

PLANNED MAINTENANCE TASK SHEET (2MD1)

The Planned Maintenance Task Sheet (2MD1) is issued for all major overhauls in conjunction with Planned Maintenance Service Sheet (1MF).

The Planned Maintenance Task Sheet will contain a very comprehensive ¯step by step¯ instruction for the necessary overhaul and also will contain the following critical information:-

> Equipment name
> Equipment number
> Criticality
> Job Number
> Craft involved
> Frequency of overhaul
> Library Manual Reference Number

It also contains a very important section concerning the safety of personnel assigned to the functions to be accomplished. The Safety section states whether a Work Permit is required and if the following equipment or tests are necessary:-

> Protective suit
> Breathing equipment
> Heating protection
> Goggles
> Gloves
> Life line
> Gas test

PLANNED MAINTENANCE SERVICE SHEET	1 M F

EQUIPMENT   DATE PREPARED   (JOB NO)

JOB FUNCTION   JOB LOCATION

WORK DESCRIPTION   CRAFT   ☐ M ☐ E ☐ I ☐ O

SAFETY INSTRUCTIONS   WORK PERMIT   PERMIT INFORMATION

YES ☐   NO

NO ☐   ISSUED BY

DURATION HRS   DATE

TOOLS REQUIRED	MATERIALS	SPECIAL TOOLS
1 2 3 4 5	1 2 3 4 5	1 2 3 4 5

SPECIAL INSTRUCTIONS   SPARES REQUIRED
1 2 3 4 5

TECHNICIAN CRAFTSMAN REMARKS   MANHOURS ☐M ☐E ☐I ☐O   SPARES USED
1 2 3 4 5

WORK COMPLETION   ■ TOTAL HOURS

SUPERVISORS REMARKS

COST CENTER

PLANNED	☐ YES	☐ NO
UN-PLANNED	☐ YES	☐ NO

DATE SCHEDULED

DATE COMPLETED

LUBRICATION CODE NO

STATUTORY INSPECTION NO

SPARES COST

MANHOUR COST

AVAILABILITY FACTOR

SUPERVISOR

DATE

48

PLANNED MAINTENANCE TASK SHEET	PAGE NO	2MD2
EQUIPMENT NO	EQUIPMENT NAME	JOB NO

TASKS TO BE PERFORMED

TASK NO		REMARKS

REMARKS

MAINTENANCE TROUBLE SHOOTING GUIDE	PAGE 3M

TYPE OF EQUIPMENT	EQUIPMENT NO

DEFECT SYMPTOM	POSSIBLE CAUSE	REQUIRED ACTION

SPECIAL INSTRUCTIONS		AREA LOCATION

DATE PREPARED

PREVENTIVE MAINTENENCE TASK SHEET | 5M(2)

EQUIPMENT NAME		JOB NO
EQUIPMENT NO	CRITICALITY	SHEET NO
CRAFT	FREQUENCY	MANUAL REFERENCE NO

SAFETY   WORK   PERMIT ☐ YES   ☐ NO

HOT WORK   ☐   GOGGLES   ☐   BREATHING EQUIP   ☐   LIFE LINE   ☐

PROTECTIVE SUIT ☐   GLOVES   ☐   HEARING PROTECTION   ☐   GAS TEST   ☐

TASK NO	TASKS TO BE PERFORMED	REMARKS

SPECIAL INSTRUCTION

DATE PREPARED	MANUFACTURER	INSTALLED

EQUIPMENT

DATE PREPARED 26.11.81 (JOB NO) AB-234

Percentage Differential Relay Protection

JOB FUNCTION

JOB LOCATION Area 1 Section 3

12 months annual Maintenance Overhaul

WORK DESCRIPTION Overhaul & Characteristic	CRAFT	☐ M ☒ E ☐ I ☐ O

SAFETY INSTRUCTIONS

	WORK PERMIT	**PERMIT INFORMATION**
Ensure unit is electrical 'Locked Out' and tagged	YES ☒	NO 443 AA 49
	NO ☐	ISSUED BY Safety Officer
	DURATION HRS 24	DATE 26.11.81 1800 hours

TOOLS REQUIRED	MATERIALS	SPECIAL TOOLS
1 Tool Box	1 Clean Rags	1 Megga
2	2 Lubricating Oil	2 Volt Ohmn Meter
3	3 Dash Pot Oil	3
4	4	4
5	5	5

SPECIAL INSTRUCTIONS	SPARES REQUIRED
Effect Characteristic Test as well as Mechanical Overhaul	1 Series 2 Relay
	2 Series 2 Spring
	3
	4
	5

TECHNICIAN CRAFTSMAN REMARKS	MANHOURS	SPARES USED
	M	1
	E	2
	I	3
		4
	O	5

WORK COMPLETION

■ TOTAL HOURS

SUPERVISORS REMARKS

COST CENTER 4493		
PLANNED	☒ YES	☐ NO
UN-PLANNED	☐ YES	☐ NO
DATE SCHEDULED 26.11.81		
DATE COMPLETED 26.11.81		
LUBRICATION CODE NO		
STATUTORY INSPECTION NO		
SPARES COST		
MANHOUR COST		
AVAILABILITY FACTOR		

SUPERVISOR

DATE

PLANNED MAINTENENCE TASK SHEET

2M

EQUIPMENT NAME		JOB NO
Percentage Differential Relay Protection		AB-234

EQUIPMENT NO	CRITICALITY	SHEET NO
330678D	(B)	1 of 2

CRAFT	FREQUENCY	MANUAL REFERENCE NO
Electrical	12 months	T.N. 58905 (46)

SAFETY   WORK PERMIT ☒ YES ☐ NO

HOT WORK	☐	GOGGLES	☐	BREATHING EQUIP	☐	LIFE LINE	☐
PROTECTIVE SUIT	☐	GLOVES	☒	HEARING PROTECTION	☐	GAS TEST	☐

TASK NO	TASKS TO BE PERFORMED	REMAR
	NOTE: Effect characteristic test as well as mechanical overhaul and inspection.	
(1)	Check Percentage Differential Element for no output signals, check the control voltage and polarity shunt the relay operation, check jacks 31 and 32 to activate the relays X124 and Y124.	
(2)	If it is found that there is signal from output relays X124 and Y124, check the coil and series resistor for disconnection. Check the final output of each printed circuit.	
(3)	If it is noted that no current flows through A/C input circuit, check the tap screws for tightness and check the drawer unit for close contact. Re-tighten all loose screws.	
(4)	High-setting over-current element for no output signals, check units for poor contact, or surface for contaminations. Also check the electric continuity of the contacts between H.O.C. relay and external terminals.	
(5)	Check unit for actuation, inspect the input. Measure the voltage between the differential voltage jacks by using a circuit tester. Approx 10V at setting valve current is considered to be the normal.	
(6)	Check Operation indicator for (A) no indication and (B) for no output signal. (A) Check the target driving indicator for specified output. Manually operate the indicator (B) Check the contacts for poor contact, also contact surface for contamination. Clean or replace.	
(7)	Check Relay Box, drawer mechanism for discolouration, deformation and deposits. Check for broken mouldings, also rubber gaskets for deformation and replace as required. Check the printed circuit board and electronic components for abnormal condition.	

SPECIAL INSTRUCTION

DATE PREPARED	MANUFACTURER	INSTALLED

PLANNED MAINTENANCE TASK SHEET	PAGE NO 2 of 2	2MD2

EQUIPMENT NO 330678D	EQUIPMENT NAME Percentage Differential Relay Protection	JOB NO AB-234

TASKS   TO   BE   PERFORMED

TASK NO		REMARKS
	CHARACTERISTIC INSPECTION	
(1)	Check operating valve. Measure operating current as follows:- (A) Setting valve and actual valve. (B) Input to single terminal (measure from the primary, secondary and terminal sides). Set 30% of setting valve current $\pm$ 8%	
(2)	Check Ration characteristic. Measure diffenential current as follows:- (A) Setting valve (differential current) actual valve. (B) Measure at 500% and 100% of outgoing current. Differential current 1^2 = 500% of setting valve current x 500% $\pm$ 10% at 500% of outgoing current	
(3)	Check Operating valve of overcurrent element as follows:- (A) Measure operating current setting valve: actual valve. (B) Input to single terminals. 900% of setting valve current $\pm$ 15%	
(4)	Check for operating time as follows:- (A) Setting valve: actual valve. (B) Input signal terminal (measure from the primary, secondary and terminal sides). (C) Measure at 300%, 500%, 1000% and 1500% of setting valve current. The percentage element within 35 m.s. at 500%. H.O.C. element within 35 m.s. at 1000%.	
(5)	Check the operating valve of operation indicator, apply rated input. It must indicate when applying the rated input.	
(6)	When necessary effect harmonic characteristic test on others to inspect for not to be operated at 15% of second harmonic.	

REMARKS

PREVENTIVE MAINTENENCE TASK SHEET | 5MC

EQUIPMENT NAME		JOB NO
Percentage Differential Relay Protection		AB-234

EQUIPMENT NO	CRITICALITY		SHEET NO
330678D	(B)		1 of 1

CRAFT	FREQUENCY	MANUAL REFERENCE NO
Electrical	3 months	T.N. 58905 (46)

SAFETY   WORK PERMIT ☒ YES   ☐ NO

HOT WORK	☐	GOGGLES	☐	BREATHING EQUIP	☐	LIFE LINE	☐
PROTECTIVE SUIT	☐	GLOVES	☒	HEARING PROTECTION	☐	GAS TEST	☐

TASK NO	TASKS TO BE PERFORMED	REMAR▶

PREVENTIVE INSPECTION

NOTE: Since this inspection is mainly relied on by the use of eyes, nose and hands, it will be difficult to find deterioration in relay performance. Pay special attention to odour, abnormal sound, discolouration and heat emission.

(1) Isolate Unit.

(2) Check contacts for discolouration, deformation and contamination. Clean or replace as necessary.

(3) Re-tighten loose tap screws and exterminal terminal screws.

(4) Check for burning odour. Check the C.T. and resistors for burning and over-heating. Withdraw and closely inspect the relay element.

(5) Check for abnormal sound, check the operation indicator output relay and H.O.C. relay for abnormal sound.

(6) Check for discolouration, check the printed circuit board output relay and H.O.C. relay interior, and tap plate also check for atmospheric gas.

(7) Check for overheating. Inspect the C.T. circuit for overcurrent.

(8) Check for signs of rust. Clean and lubricate as necessary.

(9) Check and inspect for dirt and dust on external terminals, relay box, internal unit and drawer mechanism.

(10) Inspect for cracked molding, check the external terminals, tap plate and plugs for cracked moldings.

SPECIAL INSTRUCTION

DATE PREPARED   MANUFACTURER   INSTALLED

MAINTENANCE	TROUBLE SHOOTING	GUIDE	PAGE 3M 1 of 14

TYPE OF EQUIPMENT		EQUIPMENT NO
Heavy Duty Refrigeration Compressor		

DEFECT SYMPTOM	POSSIBLE CAUSE	REQUIRED ACTION
Compressor will not start	Power off	Check main switch, fuse, wiring.
	Control power/control air off	Check switch, fuses, supply air pressure.
	Leaving gas temperature control set too high	Reset temperature setting.
	Suction damper dual pressure switch set too high	Reset pressure setting. Check if stop valve is open.
	Starter overload switch open	Reset motor overloads.
	Oil safety switch open	Reset switch.
	Dirty contacts	Clean all control contacts.
	Loose electrical connection or faulty wiring.	Tighten connections, check wiring and rewire.
	Compressor motor burned out	Check and replace if defective.
	Liquid solenoid valve closed	Check for burned holding coil, replace if defective.
	No hydrocarbon gas flow	Check gas flow through gas chiller.
	Condenser fan not in operation	Check condenser fan, fan motor fuses and overloads. Restart.

SPECIAL INSTRUCTIONS		AREA LOCATION

DATE PREPARED

MAINTENANCE	TROUBLE SHOOTING GUIDE	PAGE 3M 2 of 14

TYPE OF EQUIPMENT		EQUIPMENT NO
Heavy Duty Regrigeration Compressor		

DEFECT SYMPTOM	POSSIBLE CAUSE	REQUIRED ACTION
Compressor cycles intermittently	Low pressure switch erratic in operation	Check for clogged tubing to switch. Check switch.
	Suction throttling valve erratic in operation	Check for clogged control air tubing, control air pressure.
	Low refrigerant charge	Add refrigerant.
	Restricted liquid flow to control valves	Check flow through sight glasses. Find restriction, clean strainers. Open fully all service valves.
	Capacity control setting incorrect	Reset.
	Switch differentiation too narrow	Reset for wide load.
	Service valve on suction line not fully open	Open valve.

SPECIAL INSTRUCTIONS		AREA LOCATION

DATE PREPARED

MAINTENANCE	TROUBLE SHOOTING GUIDE	PAGE
		3M 3 of 14

TYPE OF EQUIPMENT	EQUIPMENT NO
Heavy Duty Refrigeration Compressor	

DEFECT SYMPTOM	POSSIBLE CAUSE	REQUIRED ACTION
High crankcase temperature (should be 40-42°C at seal housing)	Liquid line strainer clogged	Clean strainer.
	Excessive superheat	Reset expansion valves.
	Compression ratio too high	Recheck design.
	Discharge temperature over 125°C	Check unit application.
	Leaking suction or discharge valves	Replace valves.
	Faulty oil cooling expansion valve	Verify valve operation. Replace if faulty.
	Liquid solenoid valve does not open	Check solenoid operation. Check for clogged passage.
	Thermostat oil temperature improperly set	Reset.

SPECIAL INSTRUCTIONS		AREA LOCATION

DATE PREPARED

TYPE OF EQUIPMENT		EQUIPMENT NO
Heavy Duty Refrigeration Compressor		

DEFECT SYMPTOM	POSSIBLE CAUSE	REQUIRED ACTION
High discharge	Air in the system	Purge air.
	High expansion valve superheat	Reset expansion valves.

SPECIAL INSTRUCTIONS		AREA LOCATION

DATE PREPARED

59

TYPE OF EQUIPMENT	EQUIPMENT NO
Heavy Duty Refrigeration Compressor	

DEFECT SYMPTOM	POSSIBLE CAUSE	REQUIRED ACTION
System noises	Loose or mis-aligned coupling	Check alignment and tightness.
	Insufficient clearance between piston and valve plate	Replace defective parts.
	Motor or compressor bearings worn.	Replace bearings.
	Loose hold down bolts	Tighten bolts.
	Unit foundation improperly isolated	Isolate foundation.
	Improper support or isolation of piping	Use correct piping techniques.
	Slugging from refrigerant feedback	Check expansion valve setting. Check thermal bulb looseness and correct location.
	Hydraulic knock from excessive oil in circulation	Remove excessive oil. Check expansion valve for floodback.
	Defective valve lifter mechanism (noise level varies with un-loading)	Replace sticking filter pins. Check unloader fork for alignment. Check power element for sticking piston. Check for oil leakage at tube connection to power element. Check amount of valve pin lift above valve seat.
	Piping vibration	Support pipes as required. Check pipe connections.
	Hissing (insufficient flow through expansion valves, or clogged liquid line strainer)	Add refrigerant. Clean strainer.

SPECIAL INSTRUCTIONS		AREA LOCATION

DATE PREPARED

MAINTENANCE	TROUBLE SHOOTING GUIDE	PAGE **3M** 6 of 14

TYPE OF EQUIPMENT		EQUIPMENT NO
	Heavy Duty Refrigeration Compressor	

DEFECT SYMPTOM	POSSIBLE CAUSE	REQUIRED ACTION
Compressor will not unload	Capacity control valve not operating	Repair.
	Unloader element sticking	Repair.
	Hydraulic relay sticking	Replace control cover assembly.
	Plugged pressure line to power element	Clean line.
	External adjusting stem damaged	Replace.
Compressor will not load	Low oil pressure (below 3 BAR)	Check oil change settings.
	Capacity control valve stuck open	Repair or replace.
	Unloader element sticking	Repair.
	Plugged or broken pressure line to power element	Clean or repair.

SPECIAL INSTRUCTIONS		AREA LOCATION
DATE PREPARED		

MAINTENANCE	TROUBLE SHOOTING GUIDE	PAGE 3M 7 of 14

TYPE OF EQUIPMENT		EQUIPMENT NO
	Heavy Duty Refrigeration Compressor	

DEFECT SYMPTOM	POSSIBLE CAUSE	REQUIRED ACTION
Compressor will not unload	Capacity control valve not operating	Repair.
	Unloader element sticking	Repair.
	Hydraulic relay sticking	Replace control cover assembly.
	Plugged pressure line to power element	Clean line.
	External adjusting stem damaged	Replace.
Compressor will not load	Low oil pressure (below 3 BAR)	Check oil change settings.
	Capacity control valve stuck open	Repair or replace.
	Unloader element sticking	Repair.
	Plugged or broken pressure line to power element	Clean or repair.
	External adjusting stem damaged	Replace.
	Control oil strainer blocked	Clean or replace.
	Control valve bellows leaking	Remove thread protector and leak test. Replace valve body if bellows leaks.
	Pipe plug in pneumatic connection	Remove pipe plug.
	Foaming in crankcase from refrigerant flooding	Check expansion valve and piping.
	Hydraulic relay sticking	Replace control cover assembly.

SPECIAL INSTRUCTIONS		AREA LOCATION

DATE PREPARED

62

TYPE OF EQUIPMENT

Heavy Duty Refrigeration Compressor

EQUIPMENT NO

DEFECT SYMPTOM	POSSIBLE CAUSE	REQUIRED ACTION
Rapid unloader cycling	Excessive fluctuation in suction pressure from over-sized expansion valve	Re-size expansion valve.
	Partially plugged control oil strainer	Clean or replace strainer.
	Low oil pressure	See TROUBLE/SYMPTOM – low oil pressure.

SPECIAL INSTRUCTIONS

AREA LOCATION

DATE PREPARED

MAINTENANCE	TROUBLE	SHOOTING	GUIDE	PAGE 3M 9 of 14

TYPE OF EQUIPMENT	EQUIPMENT NO
Heavy Duty Refrigeration Compressor	

DEFECT SYMPTOM	POSSIBLE CAUSE	REQUIRED ACTION
Low oil pressure	Low oil charge	Add oil.
	Faulty oil gauge	Check and replace.
	Defective oil pressure regulator	Repair or replace.
	Clogged oil suction strainer	Clean strainer.
	Broken oil pump tank	Replace pump assembly.
	Clogged oil line	Remove obstruction.
	Worn oil pump	Replace pump assembly.
	Worn compressor bearings	Replace.

SPECIAL INSTRUCTIONS		AREA LOCATION

DATE PREPARED

| MAINTENANCE | TROUBLE | SHOOTING | GUIDE | PAGE **3M** 10 of 14 |

| TYPE OF EQUIPMENT | | EQUIPMENT NO |
| Heavy Duty Refrigeration Compressor | | |

DEFECT SYMPTOM	POSSIBLE CAUSE	REQUIRED ACTION
Cold compressor	Liquid carryover from evaporator	Check refrigerant charge and expansion valves.

| SPECIAL INSTRUCTIONS | | AREA LOCATION |

DATE PREPARED

65

TYPE OF EQUIPMENT Heavy Duty Refrigeration Compressor		EQUIPMENT NO

DEFECT SYMPTOM	POSSIBLE CAUSE	REQUIRED ACTION
Low crankcase oil	Oil return check valve stuck closed	Repair or replace check valve.
	Oil trapped in system	Check for partial oil traps in gas lines and correct piping.
	Defector oil separator float valve	Check float operation, clean blocked valve seat.
	Oil return solenoid valve does not open	Check operation of solenoid valve. Replace coil if burned out.
	Loses oil when starting – Excessive foaming	Check crankcase heater. Replace if faulty.

SPECIAL INSTRUCTIONS

AREA LOCATION

DATE PREPARED

MAINTENANCE TROUBLE SHOOTING GUIDE		PAGE **3M** 12 of 14

TYPE OF EQUIPMENT Heavy Duty Refrigeration Compressor	EQUIPMENT NO

DEFECT SYMPTOM	POSSIBLE CAUSE	REQUIRED ACTION
Cylinders and crankcase sweating	Refrigerant floodback	Check refrigerant charge and expansion valves.

SPECIAL INSTRUCTIONS		AREA LOCATION

DATE PREPARED

67

TYPE OF EQUIPMENT		EQUIPMENT NO
Heavy Duty Refrigeration Compressor		

DEFECT SYMPTOM	POSSIBLE CAUSE	REQUIRED ACTION
Compressor cycles on high pressure switch	Tubing to pressure switch blocked	Check and clean tubing.
	Faulty pressure switch	Replace or repair.
	Refrigerant overcharge	Remove excessive refrigerant.
	Insufficient flow through condenser	Check for blocked air flow to fan, clogged air passages on finned coils, loose fan belts. Remove debris, clean condenser, adjust belts.
	Discharge valve closed	Open fully discharge valve.
	Air in the system	Purge air.
	Service valve on receiver closed	Open valve.
	Liquid throttling valve closed	Check valve setting and valve operation. Reset or replace.

SPECIAL INSTRUCTIONS		AREA LOCATION

DATE PREPARED

TYPE OF EQUIPMENT		EQUIPMENT NO
Heavy Duty Refrigeration Compressor		

DEFECT SYMPTOM	POSSIBLE CAUSE	REQUIRED ACTION
Low discharge pressure	Suction service valve partially closed	Open valve.
	Leaky compressor suction valves	Open compressor and examine valve discs and valve seats. Replace if worn.
Flooding	Defective or improperly set expansion valve	Reset to 4°C Superheat. Valve operation must be stable (no hunting).
Low suction pressure	Low refrigerant charge	Add refrigerant.
	Excessive superheat	Reset expansion valves.

SPECIAL INSTRUCTIONS		AREA LOCATION

DATE PREPARED

PLANNED MAINTENANCE SERVICE SHEET   1 M F   69

EQUIPMENT	DATE PREPARED 26.11.81 (JOB NO) AP4-2344
Electronic Level Transmitter Series 12920	

JOB FUNCTION	JOB LOCATION Area 2-Section 4
12 months Maintenance Overhaul & Calibration	

WORK DESCRIPTION Complete Calibration   **CRAFT**   ☐ M ☐ E ☒ I ☐ O

SAFETY INSTRUCTIONS

Ensure unit is isolated
Inform Production

WORK PERMIT
YES ☒
NO ☐
DURATION HRS 24

PERMIT INFORMATION
NO 443AA51
ISSUED BY Safety Officer
DATE 26.11.81 1400 hours

TOOLS REQUIRED	MATERIALS	SPECIAL TOOLS
1 Tool Box	1 Clean Rags	1 Megga
2	2 Lubricating Oil	2 Volt Ohmn Meter
3	3 Grease Castrol 44	3 Mfg. Calibration Unit
4	4	4
5	5	5

SPECIAL INSTRUCTIONS

Effect Calibration Characteristic Test

SPARES REQUIRED
1 Series 2 Diaphragm
2 Series 1 Spring
3
4
5

TECHNICIAN CRAFTSMAN REMARKS

MANHOURS: ☐M ☐E ☐I ☐O

SPARES USED
1
2
3
4
5

WORK COMPLETION ■ TOTAL HOURS

SUPERVISORS REMARKS

COST CENTER	4454	
PLANNED	☒ YES	☐ NO
UN-PLANNED	☐ YES	☐ NO
DATE SCHEDULED	23.10.81	
DATE COMPLETED	23.10.81	
LUBRICATION CODE NO		
STATUTORY INSPECTION NO		
SPARES COST		
MANHOUR COST		
AVAILABILITY FACTOR		

SUPERVISOR
DATE

PLANNED	MAINTENENCE TASK SHEET	2MD1

EQUIPMENT NAME		JOB NO
	Electronic Level Transmitter Series 12920	AP4-2344

EQUIPMENT NO	CRITICALITY	SHEET NO
Spin 458890	B	1 of 4

CRAFT	FREQUENCY	MANUAL REFERENCE NO
Instrumentation	12 months Maintenance & Calibration	Fisher N70C1613

SAFETY   WORK PERMIT [x] YES   [ ] NO

HOT WORK [ ]   GOGGLES [ ]   BREATHING EQUIP [ ]   LIFE LINE [ ]

PROTECTIVE SUIT [ ]   GLOVES [ ]   HEARING PROTECTION [x]   GAS TEST [x]

TASK NO	TASKS TO BE PERFORMED	REMARKS
	REMOVAL OF AMPLIFIER	
(1)	Turn off power supply to Transmitter.	
(2)	Remove case cover.	
(3)	Remove the two nuts 12 and and pull out the Amplifier.	
	FLEXURE STRIPS AND COIL	
	To replace one or both of these elements, turn off the power supply to the transmitter, and proceed as follows:-	
(1)	Remove the case cover. Remove the screws 4 and separate the U core from the ferrite detector.	
(2)	Disconnect the coil wires from the terminal board 9.	
(3)	Loosen the wire clamp 14 and remove the Milliam Meter leads. Remove the screws 17 and remove Milliam Meter assembly to permit access to the elements behind it.	
(4)	Remove the nut 18, the load spring lever 24 and torque rod lever adapter 19.	
(5)	Loosen the two screws 21 and remove the mechanism assembly from the case.	
(6)	Loosen the square head screws 22 and remove the mechanism assembly from case.	
(7)	Remove the nut 40 and dismount the zero spring 36 from its adjustment lever 39.	

SPECIAL INSTRUCTION

Isolate unit. Remove unit to workshop.

DATE PREPARED   MANUFACTURER   INSTALLED

PLANNED MAINTENANCE SERVICE SHEET   1 M F

EQUIPMENT   DATE PREPARED 26.11.81 (JOB NO) AP4-51

Electronic Level Transmitter Series 12920

JOB FUNCTION   **JOB LOCATION** Area 2-Section 4

3 months Maintenance Inspection

WORK DESCRIPTION Inspection   **CRAFT**   ☐ M ☐ E ☒ I ☐ O

SAFETY INSTRUCTIONS

WORK PERMIT
YES ☒
NO ☐
DURATION HRS 8

PERMIT INFORMATION
NO 336B32
ISSUED BY Safety Officer
DATE 04.10.81 1200 hours

TOOLS REQUIRED
1 Tool Box
2
3
4
5

MATERIALS
1 Clean Rags
2 Lubricating Oil
3
4
5

SPECIAL TOOLS
1 Ohmn Meter
2
3
4
5

SPECIAL INSTRUCTIONS

Set Zero and check Span

SPARES REQUIRED
1
2
3
4
5

TECHNICIAN CRAFTSMAN REMARKS

MANHOURS: M ☐ E ☐ I ☐ O ☐

SPARES USED
1
2
3
4
5

WORK COMPLETION   ■ TOTAL HOURS

SUPERVISORS REMARKS

COST CENTER 4454
PLANNED ☒ YES ☐ NO
UN-PLANNED ☐ YES ☐ NO
DATE SCHEDULED 23.08.81
DATE COMPLETED 23.08.81
LUBRICATION CODE NO
STATUTORY INSPECTION NO
SPARES COST
MANHOUR COST
AVAILABILITY FACTOR

SUPERVISOR
DATE

PREVENTIVE MAINTENENCE TASK SHEET		5M(2)

EQUIPMENT NAME	JOB NO
Electronic Level Transmitter Series 12920	AP4-51

EQUIPMENT NO	CRITICALITY	SHEET NO
Spin 458890	B	1 of 1

CRAFT	FREQUENCY	MANUAL REFERENCE NO
Instrumentation	12 weeks	Fisher N70C1613

SAFETY WORK PERMIT [x] YES [ ] NO

HOT WORK	[ ]	GOGGLES	[ ]	BREATHING EQUIP	[ ]	LIFE LINE	[ ]
PROTECTIVE SUIT	[ ]	GLOVES	[ ]	HEARING PROTECTION	[x]	GAS TEST	[x]

TASK NO	TASKS TO BE PERFORMED	REMARKS
1.	Clean and remove dust on transmitter head. Set Zero and check Span.	
2.	Check and ensure smooth response throughout the Span.	

SPECIAL INSTRUCTION

DATE PREPARED	MANUFACTURER	INSTALLED

EQUIPMENT FAILURE SHEET (6M4)

EQUIPMENT NAME
Electronic Level Transmitter Series 12920

EQUIPMENT NO Spin 458890

CRITICALITY   ☐ A   ☒ B   ☐ C

LOCATION CODE   Area 2-Section 4

DATE	TIME OF FAILURE	DESCRIPTION	REPAIRS COMPLETED SPARES USED	SPARE NO	TIME ON	COMMENTS OF CRAFTSMAN	SUPERVISOR REMARKS
1. 4.8.81	1600 hours	Faulty Amplifier	Oscillated the beam by hand. Faulty ferrite	25	1630	Unit in bad state of repair. Suggest running maintenance check.	Maintenance running check brought forward by 2 months.
2. 14.8.81	1700 hours	Short circuit in Span	Replace Amplifier.	36	1725		

NON AVAILABILITY (TOTAL HOURS)   TOTAL SPARES COST   TOTAL MANHOUR COST

74

EQUIPMENT NO	EQUIPMENT NAME	JOB NO
Spin 458890	Electronic Level Transmitter Series 12920	AP4-2344

TASKS TO BE PERFORMED

TASK NO		REMARKS
(8)	Remove the five screws 30 fixing the flexure strips 32 and the zeroing springs bracket 35 on the beam 31, do not lose the spacer 33. Loosen, without removing, the four screws fixing the flexure strips on to the upper pole piece of the magnet.	
(9)	If necessary, replace the coil 27. In the case of replacement of the flexure strips, place these in position on the upper pole of the magnet.	
(10)	Re-install the beam on the mechanism, the zeroing spring bracket and its spacer, without tightening the screws. Align the beam.	
	TO REPLACE COIL	
(1)	Dismount the coil 27 by removing the two screws 26 which fix it to the beam 31 and beam extension.	
(2)	Tighten the two screws 26.	
(3)	Insert diagonally, between the upper pole piece and the 5/32 diameter a non-magnetic rod, and between the same pole piece and the beam extension a 2nd non-magnetic rod, diameter 7/32.	
(4)	While exerting a downward pressure on the beam between the alignment rods, re-tighten the flexure strips, screws 30, beginning with those on the beam.	
(5)	Remove alignment rods and tighten zeroing spring bracket screw. Connect Zeroing spring to lever 39 with zeroing spring nut 40.	
(6)	Adapt the mechanism assembly to the end of the torque tube chamber penetrating into the case, and block the square head screw 22.	
(7)	Replace knife edge block 20. Position the torque rod in centre of the torque tube with knife edge block and tighten screws 21.	
(8)	Place adaptor 19 and load spring lever on torque rod and secure with nut 19.	

REMARKS

    Isolate unit. Remove unit to workshop.

PLANNED MAINTENANCE TASK SHEET	PAGE NO 3 of 4	2MD2

EQUIPMENT NO	EQUIPMENT NAME	JOB NO
Spin 458890	Electronic Level Transmitter Series 12920	AP4-2344

TASKS TO BE PERFORMED

TASK NO		REMARKS
(9)	Fix Milliam Meter assembly on the mounting frame, fix leads under the wire clamp 14.	
(10)	Connect the coil wires to the terminal board 9 and fix the ferrite detector U core. Calibrate instrument.	
	CALIBRATION	
(1)	Connect the power supply leads to the terminal strip of the junction box, respect the polarities. Let the instrument operate for approximately 20 minutes in order to stabilize the output signal.	
(2)	Check, and if necessary adjust the transmitter action.	
(3)	Adjust process fluid to desired level for minimum output signal.	
(4)	Remove the zero adjustment cap 43 and turn the shaft 41 until the minimum signal is obtained 0% of span.	
(5)	Adjust process fluid to the desired level for maximum output signal.	
(6)	Remove the span adjustment cap 44 and turn shaft 42 until maximum signal is obtained 100% of span.	
	DISPLACER Note. One chamber type model, close stop valves.	
(1)	To remove displacer, unscrew the nuts. Remove the flange 47 and its gasket (55), and protective housing flange 53 and its gasket 54.	
(2)	Depress the torque arm 57, unhook the displacer 61 and place it in the bottom of the chamber. On the other models the displacer will be suspended within the tank. A simple hook made from 1/8'' round wire will facilitate this operation.	
(3)	Remove the screw 56 and detach the torque arm from the torque arm block 58.	
(4)	Lift the displacer from the chamber or tank.	
(5)	To replace	
	Reverse procedure for disassembly.	

REMARKS

   Isolate unit. Remove unit to workshop.

	PLANNED MAINTENANCE TASK SHEET	PAGE NO 4 of 4	2MD:
EQUIPMENT NO Spin 458890	EQUIPMENT NAME Electronic Level Transmitter Series 12920	JOB NO AP4-2344	

TASKS TO BE PERFORMED

TASK NO		REMARKS
	COVERS The case cover is screwed on to the case, and a triangular head security screw prevents its rotation. The cover is closed as follows:-	
(1)	Screw cover tightly.	
(2)	Turn back for positioning of security screw (maximum one turn).	
(3)	Block Security Screw. The junction box cover is also screwed, and its rotation is prevented by a triangular head security screw, located in the boss at the right of the case.	

REMARKS

Isolate unit. Remove unit to workshop.

MAINTENANCE	TROUBLE	SHOOTING	GUIDE	PAGE 3M

TYPE OF EQUIPMENT	EQUIPMENT NO
Electronic Level Transmitter Series 12920	Spin 458890

DEFECT SYMPTOM	POSSIBLE CAUSE	REQUIRED ACTION
1. No output current	No Electrical Supply. Faulty Amplifier.	(1) Check that the circuit is not open in the power supply or in the load if voltage exists, replace the Amplifier.
2. Less than minimum output current	Faulty Amplifier.	(2) Oscillate the beam by hand so as to move the ferrite detector from full open to full closed.
		(A) If there is no change in the current, check that a direct voltage of 14 volts $\pm$ 10% exists between the positive terminal of the terminal board 9 and the negative terminal of the power supply.
		(B) If the voltage is zero, replace the Amplifier.

SPECIAL INSTRUCTIONS		AREA LOCATION

DATE PREPARED

MAINTENANCE	TROUBLE SHOOTING GUIDE	PAGE 3M 1 of 2

TYPE OF EQUIPMENT	EQUIPMENT NO
Electronic Level Transmitter Series 12920	Spin 458890

DEFECT SYMPTOM	POSSIBLE CAUSE	REQUIRED ACTION
Less than minimum output current	Loose or open detector Faulty Amplifier	(C) If a voltage of 14 volts is present, remove amplifier and check for loose or open detector leads by connecting an Ohm Meter between socket nos. 5 and 7 on the Amplifier circuit connector. (D) If the detector is correct replace the Amplifier.
More than maximum output current	1. Faulty Coil Resistance 2. Short circuit in Span 3. Adjustment Potentiometer 4. Faulty Amplifier	(1) Oscillate the beam by hand so as to move the ferrite detector from full open to full closed. (A) If the current does not decrease, replace the Amplifier. (B) If the current passes rapidly from maximum to minimum and the reverse, and if a stabilization is not possible remove the Amplifier and check the coil resistance by connecting an Ohm Meter between the sockets nos. 4 and 8 on the Amplifier circuit connector. If the coil is interrupted (infinite resistance valve is not within the tolerance zone) replace the coil.

SPECIAL INSTRUCTIONS	AREA LOCATION

DATE PREPARED

MAINTENANCE TROUBLE SHOOTING GUIDE		PAGE 3M 1 of 2

TYPE OF EQUIPMENT Electronic Level Transmitter Series 12920	EQUIPMENT NO Spin 458890

DEFECT SYMPTOM	POSSIBLE CAUSE	REQUIRED ACTION
More than maximum output current (cont'd)		(C) If the coil circuit is correct, check that there is no short circuit in the span adjustment potentiometer. (D) If these two last controls B and C give satisfactory results, replace the Amplifier.

SPECIAL INSTRUCTIONS

AREA LOCATION

DATE PREPARED

CRITICALITY

The Criticality Code is defined as follows:-

CODE   SITUATION

(A) Vital   (1) Safety equipment whose
                                   outage creates danger or
                                   damage to platform, site
                                   or equipment, or loss of
                                   life or injury.

                              (2) Main Process equipment
                                   whose outage results in
                                   immediate production loss
                                   and penalty cost.

(B) Essential   (1) All safety equipment not
                                   already classified as
                                   vital.

                              (2) Process and auxilliary
                                   equipment whose outage
                                   does not normally cause
                                   immediate loss of produc-
                                   tion but whose continued
                                   outage (more than 24 hours)
                                   could lead to production
                                   loss or penalty.

CODE		SITUATION
(B) <u>Essential</u> (cont'd)	(3)	Life support equipment outage of more than 24 hours could cause health problems or produce dis-agreeable living conditions.
(C) <u>Support</u>	(1)	All other process equip-ment and life support equipment whose outage of more than 72 hours would affect living conditions or production.
	(2)	Commodity articles which directly support process equipment or processes. i.e. general purpose bear-ing seals, lubricants, chart paper.
(D) <u>Operational</u>	(1)	All non-industrial equip-ment and life support equipment not already classified. All commodity articles not already classified.

3   Maintenance defect monitoring

RECTIFICATION OF DEFECTS

Some means of reporting defects and requesting a service from the Maintenance Department is necessary. This is generally achieved by a Defect Card System which provides the following:-

(A)   Provides a written record of the occurence.

(B)   Ensures that the defect is reported to the Maintenance Department speedily.

(C)   Assists the Maintenance Department to ascertain priorities when compiling work programmes.

(D)   Provides a means of checking that the defect has been rectified to the satisfaction of the Production or Operating Department.

Defect Cards are usually designed in a standard pattern for different types of maintenance. Various colours being used to differentiate between Electrical, Mechanical and Instrument maintenance.

The cards are originated by the Operations or Production Departments as and when defects occur on the plant or equipment. Great care must be taken to give all relevant facts and identification of the section and plant items.

To give an indication of the urgency placed on the work the Defect Card must contain the following:-

(A)   Priority represents urgent and immediate (e.g. those defects which affect safety or availability).

(B)   Priority represents an important job which requires attention as soon as possible.

(C)   Priority represents a less important and minor job (e.g. those defects which can wait until labour is available).

The following further information to assist in work planning is also desirable:-

(1)   Necessity for a `Permit to Work`.

(2)   Necessity for a unit outage or any special isolation arrangements.

Routing of Cards and Operation of System

(1)   Cards are made out in triplicate, consisting of two copies and a card. One copy is retained by the originator.

(2)   The Card and one copy arrive at the Maintenance Office and are sorted. Priority cards are passed immediately to Departmental Supervisors. Supervisor examines Defect Cards and controls flow to Maintenance personnel. Cards not requiring immediate attention are filed `Awaiting attention`.

(3)   Foreman allocates card to craftsman and copy filed in `Job in Hand Rack`.

(4)   Craftsman completes job, returns card endorsed with necessary information.

(5)   Copy removed, cancelled and returned to originator.

(6)   Maintenance information on card is checked by foreman and passed to Department Engineer, who indicates the items to be recorded in plant history files.

(7)   Card is finally filed and kept for reference for possibly two years.

DEFECT CARD OPERATION DEPARTMENT		NO. 231
LOCATION (B) Area EQUIPMENT NAME Steam Turbine Condenser		
PRIORITY A ☐ B ☐ C ☐ SECTION Power Generation	A UNIT	
PERMIT TO WORK YES ☐ NO ☐ EQUIPMENT LOCKOUT YES ☒ NO ☐		
ISOLATION REQUIREMENTS Suction & Delivery Valves to be locked out		

NATURE OF DEFECT

(B)   Feed Water Regulating Valve Gland leaking

REPORTED BY NAME: Assistant Charge Engineer   SIGNATURE
DATE: 25.7.81
TIME: 1600 Hours

ROUTE OF MECHANICAL DEFECT CARDS

MANAGEMENT
OPERATIONS
PRODUCTION

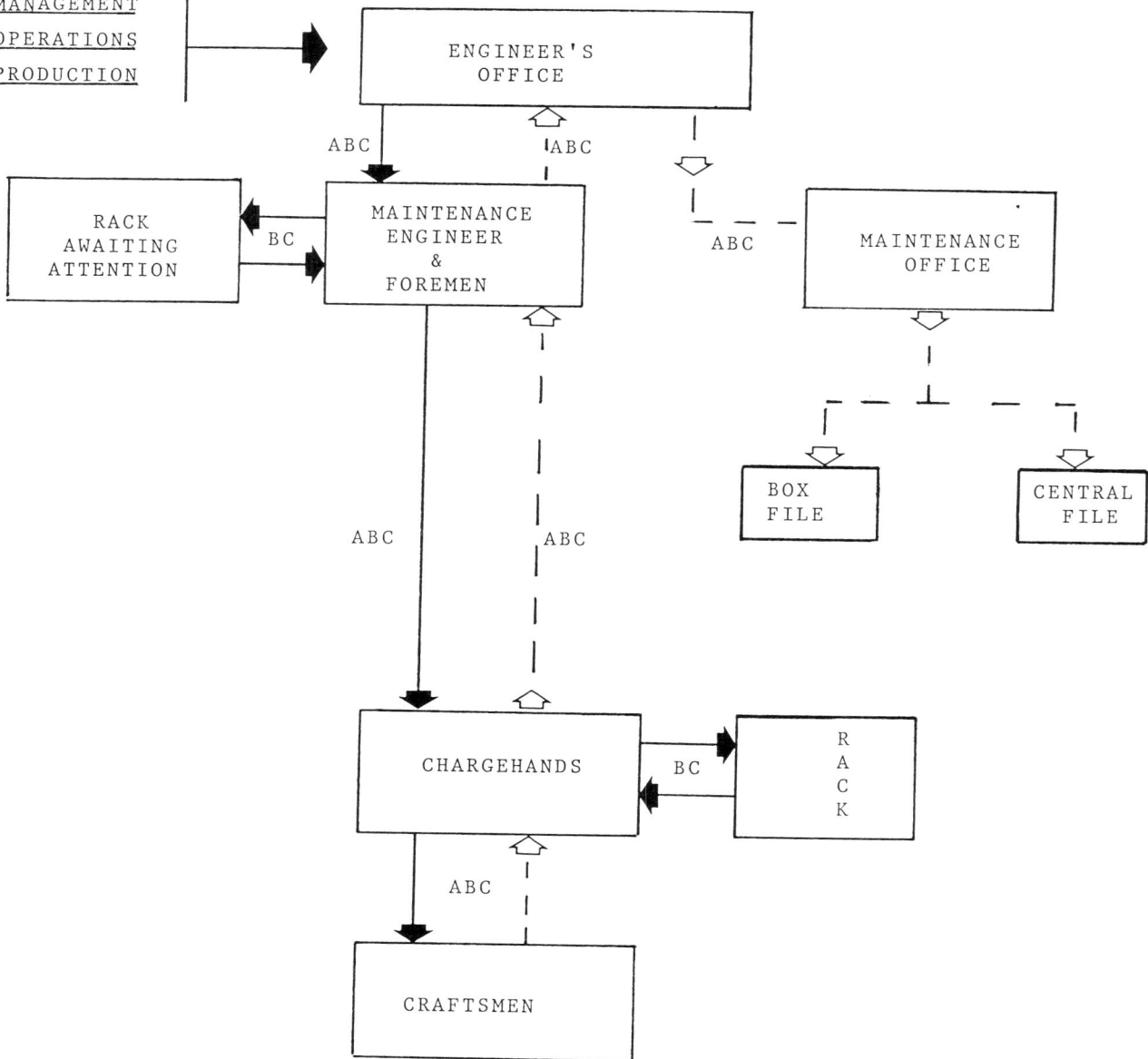

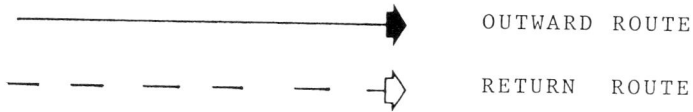

ENGINEER'S OFFICE

ABC

ABC

ABC

RACK AWAITING ATTENTION

BC

MAINTENANCE ENGINEER & FOREMEN

MAINTENANCE OFFICE

ABC

BOX FILE

CENTRAL FILE

ABC

ABC

CHARGEHANDS

BC

RACK

ABC

CRAFTSMEN

OUTWARD ROUTE

RETURN ROUTE

LETTERS INDICATE PRIORITY

FEEDBACK

(1.1) As in any system, the feedback or transmitting of information is indispensable. This does not only apply to the tracking of the defect by the Maintenance Department, but also to the Operations Department.

A document was designed not only to fulfil the Maintenance Department's needs, but also those of the Operations Department.

This document will be utilised by the Planning Craft Sections to plan and instigate required defect rectifications over a given period. The defect will be recorded by the following:-

     (A)   Defect No.

     (B)   Date Defect Raised

     (C)   Defect Originator

     (D)   Defect Priority

     (E)   Defect Craft

     (F)   Defect Description

(1.2) Defect (S) = shutdown defect will be recorded and issued as per the planned maintenance outage programme as applicable to the plant equipment.

A situation could occur whereby the said defect would be brought forward or revised due to a planned unit shut-down for reasons other than a planned maintenance outage programme, ie: operation non-requirement of a production plant, or multiple defect accumulation due to a singular piece of equipment.

It will then be the responsibility of the Planning Department to issue the applicable defects to the craft sections as and when a unit is removed from operational service. All defect issued will be subject to shut-down unit duration.

(1.3) Defect (B) = B priority. B priority defects will be recorded and issued according to a given priority subject to materials and spare parts, and available manpower. Each defect will be allocated a given week within the year for required rectification subject to priority.

E EMERGENCY DEFECT

Outside Normal Working Hours

In the event of an E emergency defect occurring after the normal working hours, direct communication is required with the appropriate Maintenance Section Head, or if not available, then the Area Engineer. This avenue of direct communication is most important as speed is essential. It will be the responsibility of the above named persons to ascertain the situation in liaison with the Shift Engineer and arrange any necessary call out for the affected personnel. Clarification of the situation can be made to the Maintenance or Operations Superintendents if necessary.

The Shift Engineer will prepare permit-to-work documentation, if possible, before the arrival of the maintenance emergency team.

Normal Working Hours

In the event of an E emergency defect occuring during working hours, the following procedure will be instigated. Direct communication will be made with the Maintenance Section Head. He will ascertain the situation in liaison with the Shift Engineer and appoint required maintenance personnel. Clarification applicable to the situation can be made to the Maintenance or Operations Superintendents as necessary.

As above, the Shift Engineer will prepare the required permit-to work documentation as soon as possible.

NORMAL WORKING HOURS: FLOW CHART 1

(1.1)   All defects raised during the normal working hours
        except in the case of emergency defects will be
        collected from the shift control room by the job preparer.

(1.2)   The Planning Section Head will discuss priority A defect
        cards with appropriate Maintenance Section Heads, which
        will include duration, priority, overtime requirements,
        etc.

(1.3)   Overtime authorization for any A considered defects will
        be considered by the Maintenance Superintendent in liaison
        with the Operations Superintendent. All non-considered
        overtime defects will either commence the same day with
        carry over, or be scheduled the following working day.

(1.4)   The defect procedure applicable to defects B and S will be
        subject to the same procedure outside normal working
        hours as mentioned on p. 87.

(1.5)   Procedure applicable to E emergency defects will be as
        described earlier on p. 83.

(1.6)   Considered or agreed A defect white and green copies to
        submitted direct to the appropriate Maintenance Section
        Head for required instigation. Pink copy retained by
        Planning Section Head.

        Non-considered A defects returned to shift engineer for
        reclassification. Reclassified A defect will follow the
        normal route applicable to B or S.

(1.7)   The Planning Section Head will transport all B and S
        defect cards including pink copy of A defects to the
        central planning coordination and records office. The
        defect cards will be registered as A, B, or S priority
        by the job preparers. B defect cards will be issued
        to the Maintenance Section Heads according to priorities
        and manpower availability. The pink copy to be retained
        by the Planning Department for tracking and progress.

The green copy to be issued to the appropriate Maintenance Section Head for required instigation.

(1.8)   All S shut-down defect cards will be recorded and issued as per the plant planned maintenance outage or plant/unit shut-down, the latter being subject to outage duration. A copy of the shut-down defect to be issued to appropriate Maintenance Section Head for instigation, and a pink copy retained. It will be the responsibility of the Planning Department to file by plant/equipment name all S type defects for immediate distribution prior to any plant/ equipment outage.

(1.9)   All returning rectified defects A, B, S, and E.1 will be monitored and recorded, ie: duration of rectification, manhours applicable, spare parts, etc. Planning and Scheduling Engineer will conduct investigation into accumulative defects, recommend action to elevate said accumulation, and issue defect priorities after liaison with Maintenance Section Heads/Job Preparers. The defect serial number will be forwarded to originator/ shift engineer on completion of defect, ie: A, B, S, and E.1.

(1.10) Planning and Scheduling Engineer to modify planned/ preventive maintenance system/schedule to elevate accumulative defects if possible.

(1.11) The following items will be completed by the shift engineer on the defect card after it has been established that the defect has not been previously reported:

(1)   Location of the defect.

(2)   Description of defect.

(3)   Appropriate discipline if possible.

(4)   Allocate priority, i.e. A, B, S, or E.

(5)   Allocate defect number (serially).

DEFECT PROCEDURE (1) FLOW CHART (1)

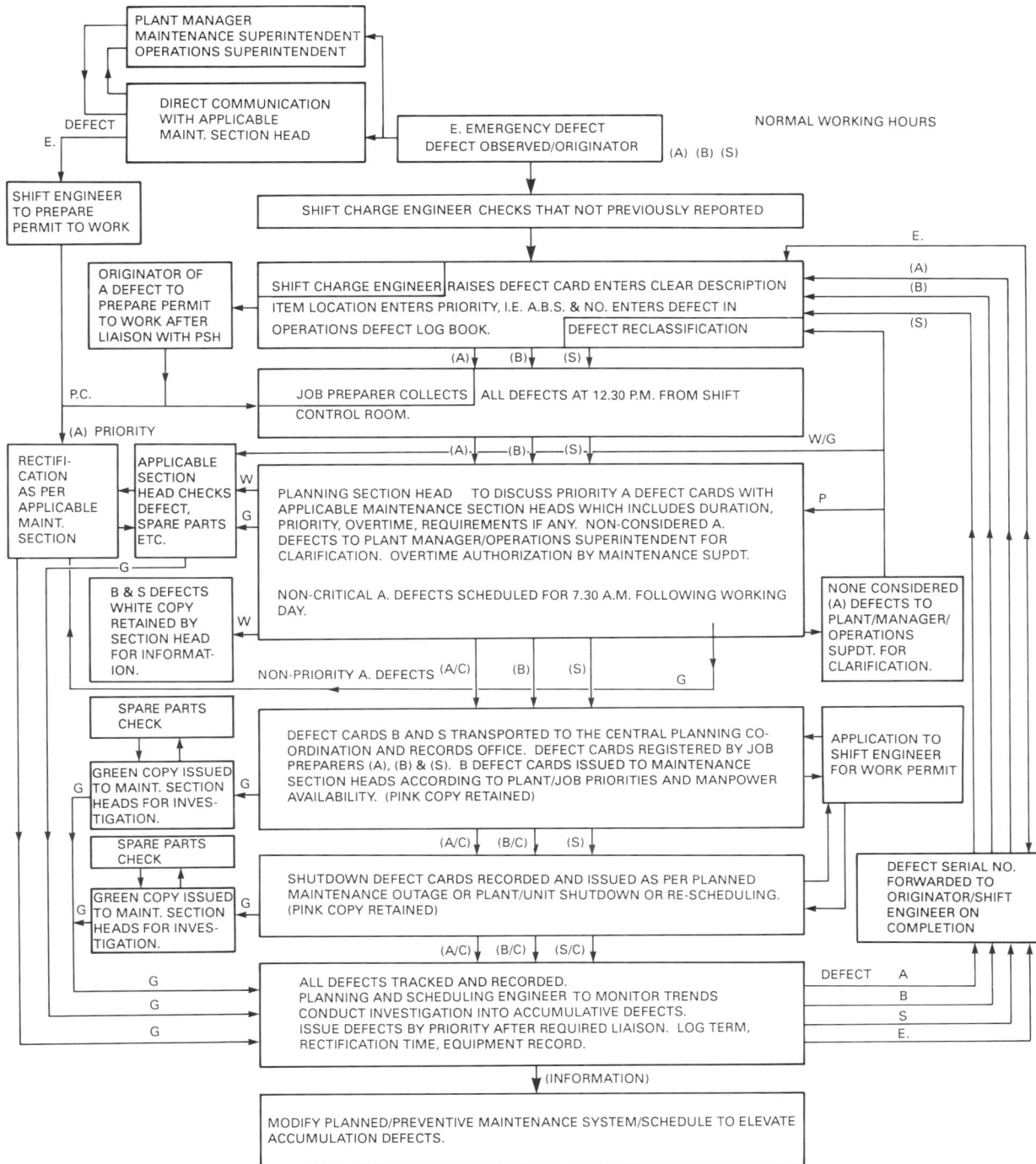

NORMAL WORKING HOURS

NOTE:
(1) DEFECT PRIORITY
    (A) = A PRIORITY   (B) = B PRIORITY   (S) = SHUTDOWN/OUTAGE
(2) DEFECT CARD
    THE DEFECT CARD COMPRISES OF 3 DIFFERENT COLOURS
    W = WHITE   P = PINK   G = GREEN
(3) E = EMERGENCY DEFECT

OUTSIDE NORMAL WORKING HOURS: FLOW CHART 2

(1.1)   It will be the responsibility of the shift engineers to prepare and have available the required permit to carry out work on the A priority defects. This work permit should be available early morning for collection by the Planning Section Head.

(1.2)   The Planning Section Head will collect all defect cards as raised from the shift engineer in the morning. The defects will include A, B, and S. (E defect card flow chart is described on p. 83.) A considered defects will be accompanied by the applicable work permit.

(1.3)   On receipt of the above mentioned defects the Planning Section Head, after liaison with the appropriate Maintenance Section Heads, will select the A defects as a priority for immediate distribution. The green and white copies are distributed, with the pink copy retained by the Planning Department.

(1.4)   The Planning Section Head and Maintenance Section Heads will attend the revised maintenance morning meeting to discuss all defects. B and S defect white copies to be issued to appropriate section heads for information. The Planning Section Head will retaine the B and S defect green and pink copies for future distribution.

Non-considered A defects will be discussed by the Plant manager/Operations Superintendent/Maintenance Superintendent for clarification at a later meeting.

8

DEFECT PROCEDURE FLOW CHART (2)

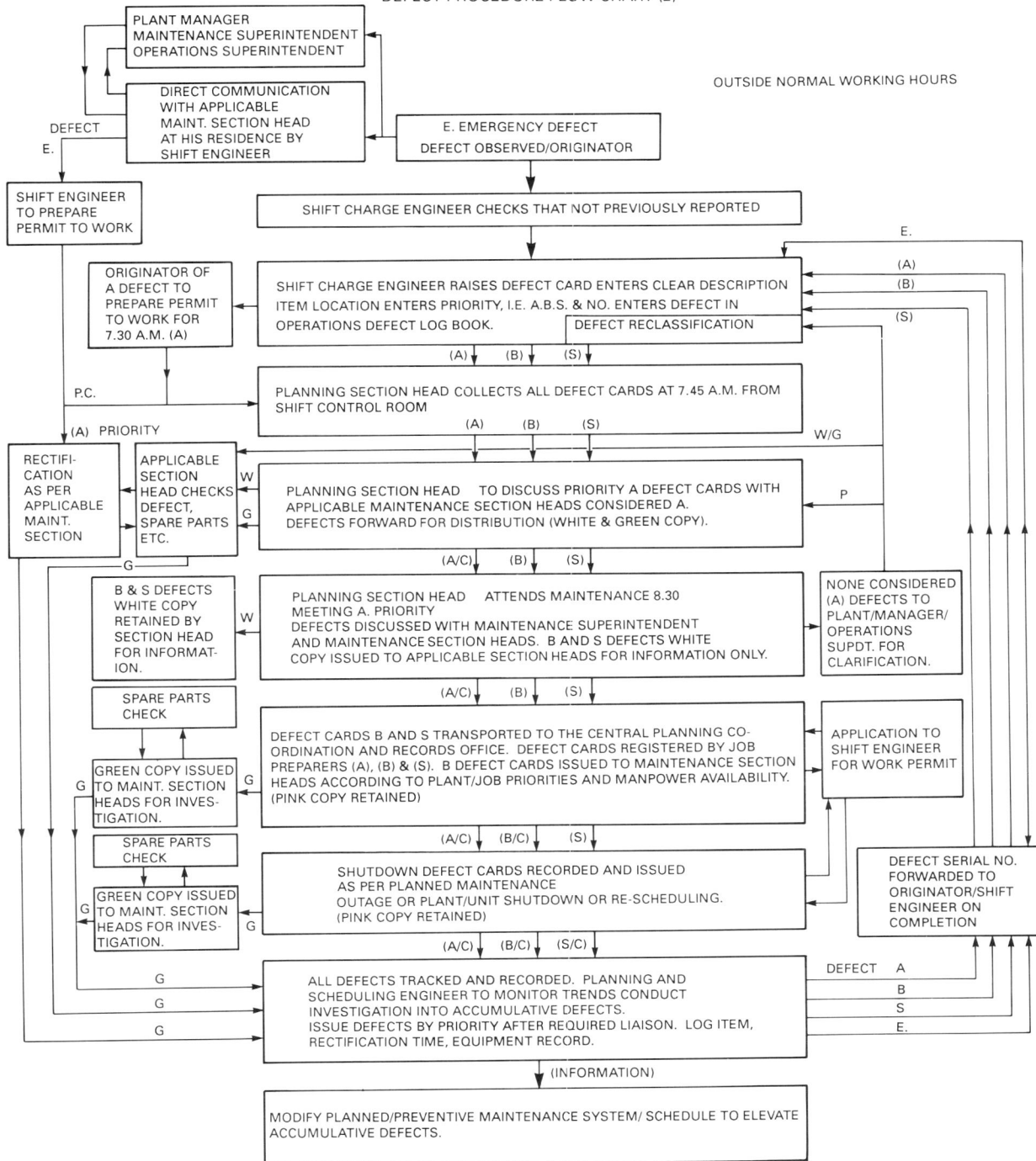

NOTE:
(1) DEFECT PRIORITY
    (A) = A PRIORITY   (B) = B PRIORITY   (S) = SHUTDOWN/OUTAGE
(2) DEFECT CARD
    THE DEFECT CARD COMPRISES OF 3 DIFFERENT COLOURS
    W = WHITE   P = PINK   G = GREEN
(3) E = EMERGENCY DEFECT

96

OBJECTIVES

(1.1)   To enable the Operations Department to notify the occurrence of any plant or equipment defect requiring the attention of the Maintenance Department.

(1.2)   To provide the Maintenance Department with a concise and accurate means of monitoring the organizational flow and preparations necessary to instigate the rectifications.

(1.3)   To provide a means of transmitting information which is required to be entered on the plant/equipment record cards (History).

(1.4)   To provide required feedback to the Maintenance Department on defect accumulation or repetition.

(1.5)   To allow notification to the Operations Department that the requested defect work has been completed.

(1.6)   To allow the Maintenance Department to instigate any modifications to the planned/preventive maintenance system or otherwise to elevate repetitive defects.

ROUTING

(1.1)   The routing of defect reports will be in accordance with Flow Charts 1 and 2 (pp. 86 & 88). All defect reports will be processed by the Planning Section in conjunction with the appropriate Maintenance Section (craft).

PROCEDURE

(1.1) The defect form comprises of three different coloured
       sheets, ie: white, pink and green. Pads of the said
       defect forms will be retained in the Operations control
       room. The form will normally be authorized by the shift
       engineer. However, engineers and above have the right to
       initiate a defect. For control purposes all authorized
       defects should be recorded through the shift engineer.

(1.2) All defects, ie: E, A, B, and S will be tracked and
       recorded. The Planning and Scheduling Engineer will
       monitor trends, conduct investigations into accumulative
       defects, and issue defects by priority after required
       liaison. The defect serial number will be forwarded to
       the originator/shift engineer on completion.

WATER & POWER STATION (RUWAIS)
MAINTENANCE DEPARTMENT

PAGE ONE

98

Accumulative / Repetition Defects

DEFECT HISTORY

Defect Item No	Defect NO	Priority	Date Raised	Craft	Decription	Defect	Month	198

E A B S M E O

Item →
Craft →

WEEK → 1 2 3 4 5 6 7 8 9 10 11 12 13 14 15 16 17 18 19 20 21 22 23 24 25 26 27 28 29 30 31 32 33 34 35 36 37 38 39 40 41 42 43 44 45 46 47 48 49 50 51

SPARE PARTS   Originator

Hours D   Completion D

JAN FEB MAR APR MAY JUN JUL AUG SEPT OCT NOV DEC

REMARKS

◣ DEFECT SCHEDULED

☐ DEFECT COMPLETED

◸ DEFECT RESCHEDULED

WATER & POWER STATION (RUWAIS)
MAINTENANCE DEPARTMENT

PAGE ONE

DEFECT HISTORY   Month JUNE   1986

Accumulative / Repetition Defects

Defect Item No	Defect NO	Priority	Item	Description	Spare Parts	Originator / Completion D	Hours / Remarks
1 / 2	7237	S M	10	TAPROGGE SYSTEM	CD	AVAIL	
3	7238	S M B	10	FUEL OIL FILT. GLAND	SG	AVAIL	11.6.86 6 HRS
4	7239	M A	11	FERTIL PUMP A	PD		14.6.86 4 HRS
5	7240	E B	11	C.C.R. LIGHTS FUSED	SU		13.6.86 5 HRS
6	7241	M I A	11	POWER PUMP B	CD	SG	
7	7242	I B	12	CHLORINE BANK	PD		
8	7243	I B	12	F.W. FLOW TANK	CD		
9	7244	M E A	12	BAND SCREEN (A)	CD		
10	7245	E B	12	CHL/SY. EVAP. B	SU	AVAIL	
11	7246	M S	12	B FEED WATER	SG		13.6.86 6 HRS
12	7247	E E	13	33 KV S/G	PD		13.6.86 3 HRS
13	7248	E E	13	TRANS. LINE		PROG	
14	7249	E S	14	F.D. FAN UNIT (1)	CD	PROG	14.6.86 8 HRS
15	7250	M S A	14	WATER LINE (H)	CD		
16	7251	M E A	16	BAND SCREEN (A)	CD		16.6.86 8 HRS
17	7252	M I A	16	POWER PUMP (B)	PD		
18	7253	S M	16	CHLORINE (BANK)	PD	PROG	
19	7254	S I E	16	F.D. FAN UNIT 2	PD	PROG	
20	7255	S E	16	F.D. FAN UNIT 2	PD		

WEEK → 1 2 3 4 5 6 7 8 9 10 11 12 13 14 15 16 17 18 19 20 21 22 23 24 25 26 27 28 29 30 31 32 33 34 35 36 37 38 39 40 41 42 43 44 45 46 47 48 49 50 51

JAN   FEB   MAR   APR   MAY   JUN   JUL   AUG   SEPT   OCT   NOV   DEC

REMARKS

◣ DEFECT SCHEDULED
■ DEFECT COMPLETED
◤ DEFECT RESCHEDULED

99

MAINTENANCE DEPARTMENT

INSTRUMENTATION SECTION

DEFECT ACCUMULATION WITH APPLICABLE MANHOURS FOR REQUIRED RECTIFICATION (1985)

12TH JANUARY, 1986

PRIORITY A & B

	DEFECT CARDS NUMBERS/MANHOURS				
MONTH	PRIORITY (A)	PRIORITY (B)	TOTAL NUMBERS	PRIORITY (A) MANHOURS	PRIORITY (B) MANHOURS
JANUARY	27	79	106	62	190
FEBRUARY	24	52	76	81	149
MARCH	33	79	112	67	171
APRIL	31	73	104	109	236
MAY	50	54	104	153	118
JUNE	31	88	119	98	155
JULY	58	63	121	214	131
AUGUST	31	87	118	77	242
SEPTEMBER	38	59	97	181	144
OCTOBER	40	54	94	174	187
NOVEMBER	26	41	67	74	89
DECEMBER	27	79	106	75	176
TOTAL	416	808	1224	1365	1988

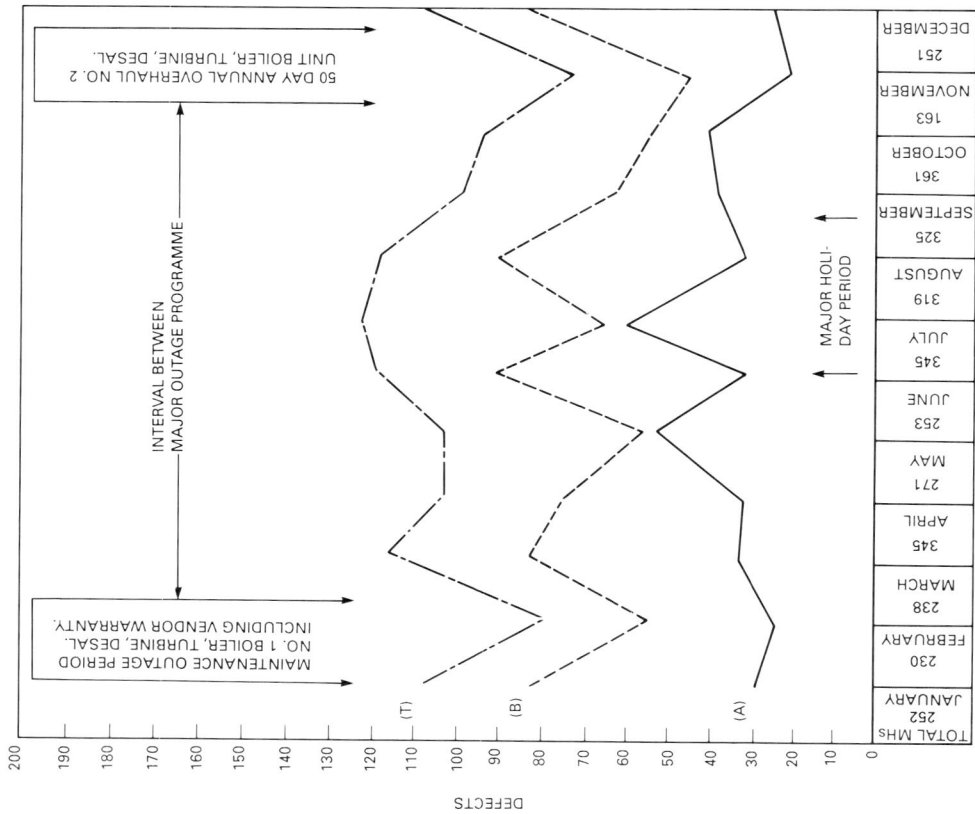

DEFECTS

MAINTENANCE OUTAGE PERIOD NO. 1 BOILER, TURBINE, DESAL. INCLUDING VENDOR WARRANTY.

INTERVAL BETWEEN MAJOR OUTAGE PROGRAMME

50 DAY ANNUAL OVERHAUL NO. 2 UNIT BOILER, TURBINE, DESAL.

MAJOR HOLI-DAY PERIOD

(T)
(B)
(A)

1985

| TOTAL MHS | JANUARY 252 | FEBRUARY 230 | MARCH 238 | APRIL 345 | MAY 271 | JUNE 253 | JULY 345 | AUGUST 319 | SEPTEMBER 325 | OCTOBER 361 | NOVEMBER 163 | DECEMBER 251 |

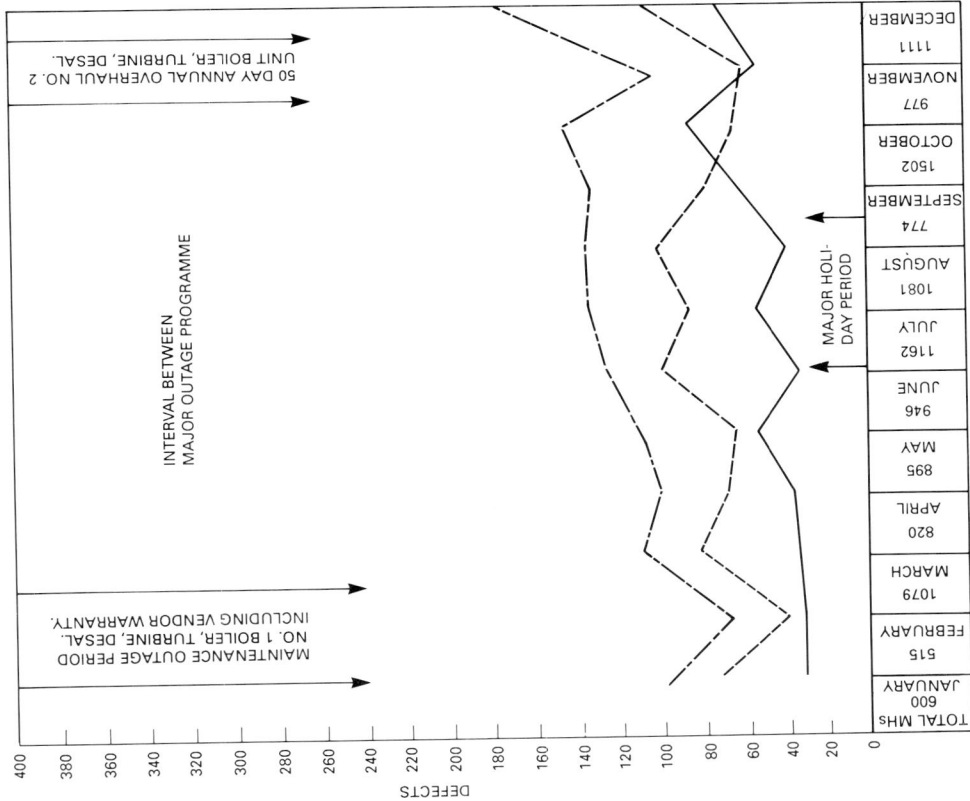

101

MAINTENANCE DEPARTMENT

MECHANICAL SECTION

DEFECT ACCUMULATION WITH APPLICABLE MANHOURS FOR REQUIRED RECTIFICATION (1985)

PRIORITY A & B

| | DEFECT CARDS NUMBERS MANHOURS | | | | | |
MONTH	PRIORITY (A)	PRIORITY (B)	TOTAL NUMBERS	PRIORITY (A) MANHOURS	PRIORITY (B) MANHOURS
JANUARY	28	68	96	244	356
FEBRUARY	28	43	71	218	297
MARCH	30	82	112	531	548
APRIL	35	64	99	414	406
MAY	48	61	109	485	410
JUNE	33	91	124	319	627
JULY	48	79	127	527	635
AUGUST	37	94	131	293	788
SEPTEMBER	58	68	126	462	312
OCTOBER	84	59	143	1037	465
NOVEMBER	48	53	101	537	440
DECEMBER	70	104	174	497	614
TOTAL	547	866	1503	5564	5898

NOTE: WPS has no work order/requests for service as applicable to other HP sites. All services to Operations are logged on a 'Defect Card', i.e. 'chlorine cylinder change out', etc.

MAINTENANCE DEPARTMENT

DEFECT ACCUMULATION WITH APPLICABLE MANHOURS FOR REQUIRED RECTIFICATION (1985)

| | DEFECT CARDS NUMBERS MANHOURS | | | | |
MONTH	PRIORITY (A)	PRIORITY (B)	TOTAL NUMBERS	PRIORITY (A) MANHOURS	PRIORITY (B) MANHOURS
JANUARY	59	160	219	314	603
FEBRUARY	63	110	173	334	516
MARCH	70	191	261	644	847
APRIL	80	163	243	597	769
MAY	104	128	232	667	588
JUNE	70	193	263	440	839
JULY	110	167	277	790	878
AUGUST	77	213	290	406	1177
SEPTEMBER	105	148	253	717	612
OCTOBER	135	139	274	1268	769
NOVEMBER	86	108	194	709	636
DECEMBER	110	212	322	644	960
TOTAL	1069	1932	3001	7530	9194

NOTE: WPS has no work order/requests for service as applicable to other HP sites. All services to Operations are logged on a Defect Card, i.e. 'Chlorine cylinder change out', etc.

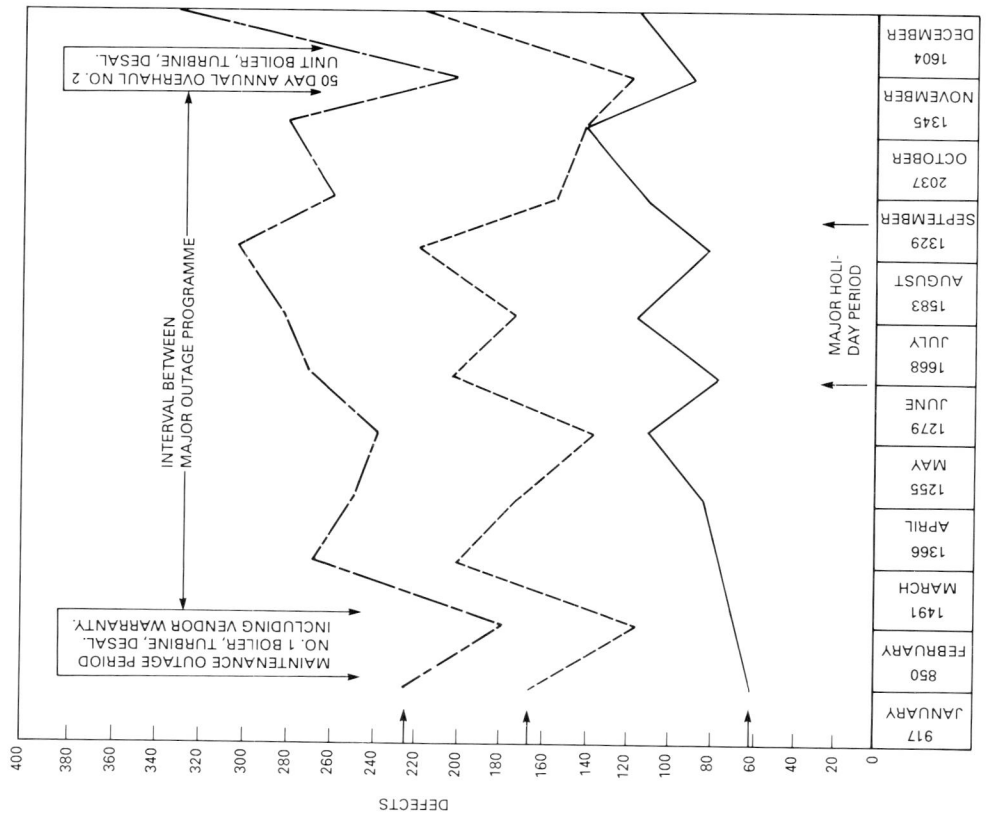

PRIORITY A & B

DEFECTS

INTERVAL BETWEEN MAJOR OUTAGE PROGRAMME

50 DAY ANNUAL OVERHAUL NO. 2 UNIT BOILER, TURBINE, DESAL.

MAINTENANCE OUTAGE PERIOD NO. 1 BOILER, TURBINE, DESAL. INCLUDING VENDOR WARRANTY.

MAJOR HOLI-DAY PERIOD

Month	Value
JANUARY	917
FEBRUARY	850
MARCH	1491
APRIL	1366
MAY	1255
JUNE	1279
JULY	1668
AUGUST	1583
SEPTEMBER	1329
OCTOBER	2037
NOVEMBER	1345
DECEMBER	1604

4 Maintenance stores spare part monitoring

STORES

STORES STOCKS - Identification and Classification

Maintenance stores stocked in the petro/chemical industry
are broadly of two types.

(1)   General and Engineering stores - these are items of
       general use such as nuts, bolts and screws, small
       machine and hand tools, raw metals, jointing and
       packing, which are normally readily obtainable from
       a number of suppliers. Cleaning materials, protec-
       tive clothing, electric light fittings, lamps and
       other domestic-type materials are also included in
       this category.

(2)   Plant Spares - these are associated with a particular
       item of plant or equipment and the source of supply
       is in most cases limited to the original manufacturer
       of the plant. A creep-resisting bolt and nut for a
       turbine flange for example would be treated as a
       plant's spare rather than a general item of stores.

       All items should be coded within the framework of the
       company's commodity code. There should be a seven
       digit code - the application of which is demonstrated
       by the following example:-

       Code = 0237047

       02 - Main Group Class - Fastenings and Fixings

       37 - Sub Group Class - Screws, Black Low Carbon (BSW)

       047 - Individual Item Detail - Hexagon Head 1/4 X 3/4
                                                  Long

This method provides a positive means of identification.
Overall control of stocks is maintained by a materials
control card index system, one card for each item, which
contains full information on:-

    (1)   Receipts
    (2)   issues
    (3)   physical stock held
    (4)   orders placed
    (5)   bin location
    (6)   re-order level min/max

and in addition to the code number a detailed description
of the item for re-ordering purposes.

These records are checked on a perpetual inventory basis
against the physical stock and reconciled with the stores
ledger account. They are also used in connection with a
system of stock review to highlight for disposal any items
which have become obsolete or redundant.

ILLUSTRATION OF SPARES STOCK

Spares Code	Item No.	Part No.	Brief Description	Maker's Catalogue		Spare part class	Quantity Per Engine	Recommended Quantity	Re-order Level
61.135	55	BA.33414	Bush	01	02	1	1	2	2 R.A.U.
137	57	KB.7130	Tab Washer	01	02	5	1	25	13
138	58	U.122314	Nut	01	02	1	1	25	7
146	59	KB.16726	Sealing Ring	01	04	5	2	5	2
161	60	BA.31875	Screw	01	08	1	1	3	3 R.A.U.
162	61	BA.32757	Washer	01	08	5	1	20	10
163	62	BA.32758	Element	01	08	1	1	1	1 R.A.U.
164	63	BA.32759	Element	01	08	1	1	1	1 R.A.U.
165	64	BA.80956	Joint	01	08	5	1	25	13
166	65	BA.80969	Joint	01	08	5	1	25	13
167	66	BA.80970	Joint	01	08	5	1	12	6
168	67	BA.80971	Joint	01	08	5	1	25	13
169	68	BA.80972	Joint	01	08	5	1	12	6
170	69	BA.80973	Joint	01	08	5	1	12	6
171	70	BA.80975	Joint	01	08	5	1	12	6
172	71	BA.80980	Joint	01	08	5	1	12	6
173	72	BA.80982	Joint	01	08	5	2	25	13
174	73	BA.93287	Joint	01	08	5	1	12	6
175	74	BA.99822	Sealing Ring	01	08	5	1	5	3
176	75	BA.99826	Sealing Ring	01	08	5	1	20	10
177	76	BR.42344	Joint	01	08	5	1	12	6
178	77	BR.47376	Screw	01	08	1	1	1	1 R.A.U.
179	78	CU.3895	Joint	01	08	5	1	6	3
180	79	K.4556	Tab Washer	01	08	5	10	125	63
181	80	KB.14162	Circlip	01	08	5	1	5	3
182	81	KB.16501	Sealing Ring	01	08	5	5	25	13
183	82	KB.16502	Sealing Ring	01	08	5	5	20	10

108

COMPUTERISED STORES SYSTEM
FLOW SHEET

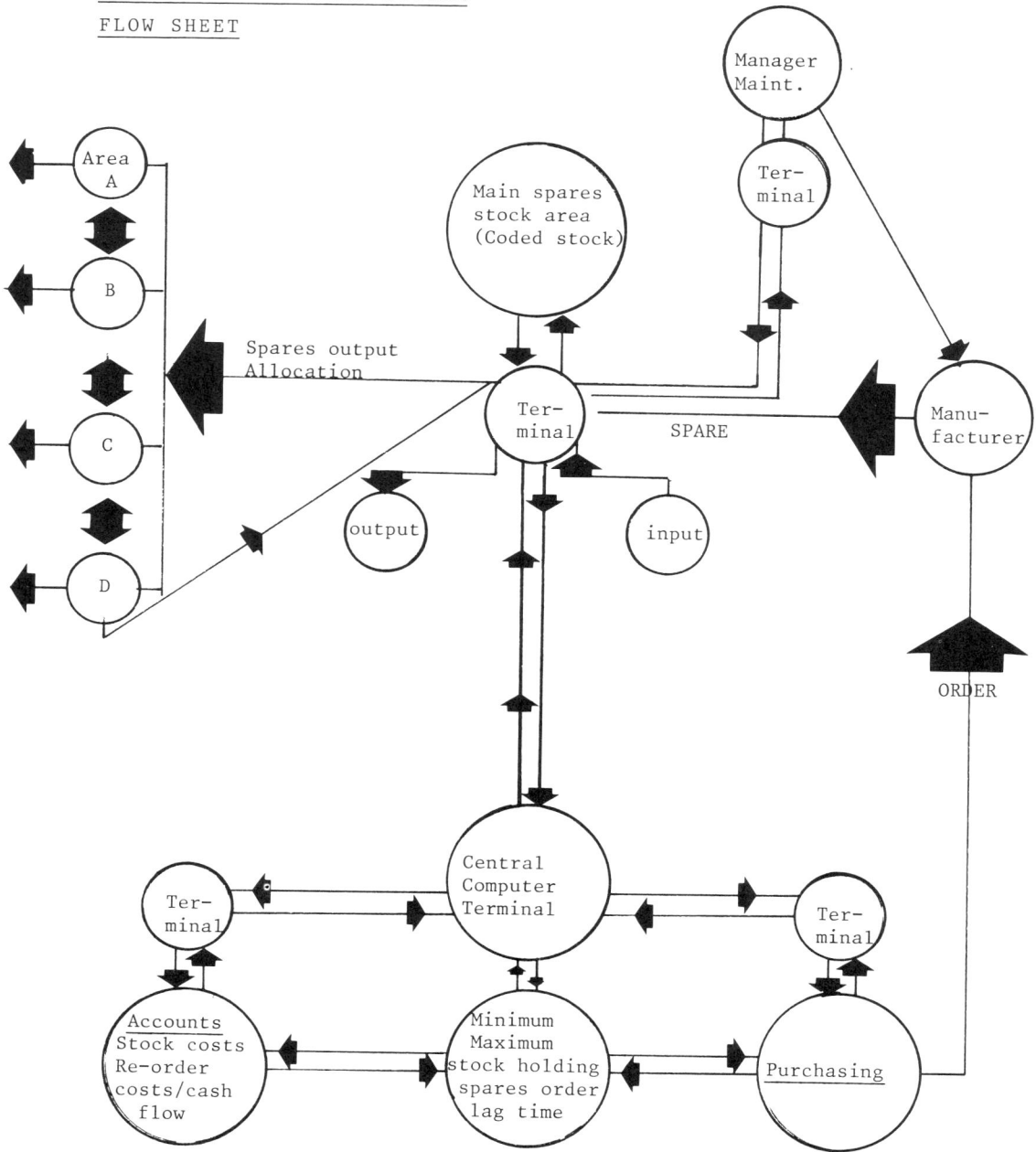

STOCK LEVELS

Stock levels for items with a predictable rate of use are determined on a basis which takes into account expected consumption, overall delivery (time taken from re-ordering and delivery on site), price of the item and the administrative costs.

The aim is to keep stocks at a minimum economical level consistent with the needs of repair and maintenance work (Planned and Preventive), also feed-back received from the Planned Maintenance Service Sheet No. 1MF

In determining stock levels, account must be taken of high comsumption over a short period such as may arise at the time of a major plant overhaul. Excess stocks are prevented by the preparation of lists of stores items required for an overhaul at the planning stage (Planned Maintenance Service Sheet 1MF), long time stocking of spares would also be informed by accumulated information received from Planned and Preventive Maintenance feed-back and information received from the Failure Sheets.

5   Statutory inspection monitoring

<u>STATUTORY RECORDS</u>

In addition to the "Equipment Record cards" being filed on each independent piece of plant and equipment, certain Statutory Records must also be filed and they normally consist of the following items:

(A)   <u>PRESSURE VESSEL RECORDS</u>

All pressure vessels must be examined at prescribed periods by a competent person and the results of such examinations attached or kept near to the General Register. Pressure vessel examinations are usually carried out by surveyors employed by the applicable government or insuring authority and certificates of the examinations are filed by the Maintenance Section.

The plant items for examination are listed on schedules which are normally issued by the applicable authority following the surveyor's preliminary examination. Details of any new plant installed and commissioned must be forwarded to the applicable authority for inclusion in the inspection schedule.

Pressure Vessels normally covered by Statutory Inspection fall into three main groups:

(1)   <u>Steam boilers</u> - this group includes water tube and drum boilers, economisers, superheaters and reheaters.

(2)   <u>Steam receivers and steam containers</u> - included in this group are feed heaters, deaerators, desuperheaters, drain coolers and similar vessels.

(3)   <u>Air receivers</u> - this group includes associated after coolers, inter coolers and pressure-type filters.

(B)   LIFTING TACKLE RECORDS

Every production organisation has many items of lifting tackle which are normally included in any Statutory Inspection. In order to comply with any inspection scheme, a rigid system of recording is essential.

Lifting equipment normally covered by this section include the following:

(1)   Cranes
(2)   Runway beams
(3)   Trolleys
(4)   Chain blocks
(5)   Pull lifts, rope blocks and rotor lifts
(6)   Eye bolts
(7)   Shackles and hooks
(8)   Wire rope slings
(9)   Chain slings
(10)   Turbine lifting gear and special rigs

Cards are made out in the appropriate section for each item, and are numbered with the S.P.I.N. & F.I.N. number (stamped on the equipment) prefixed by the letter for its section. In the case of wire ropes associated with an item of lifting equipment e.g. cranes - the card for the rope is given the same number as the crane with which it is associated and filed in the crane section.

Each card contains such information as:

(1)   Manufacturer's name
(2)   Name and address of the person who issued the test
       certificate
(3)   Test certificate number
(4)   Safe working load
(5)   Commission date of equipment

Some of the above information would be entered on the Equipment Record card, while the remainder would be attached on the reverse side of the above mentioned card.

LIFTING EQUIPMENT/TEST PROGRAMME AND SCHEDULE CHART

The Lifting Equipment test programme chart should allow the Maintenance Manager or his staff to see at a glance the current situation regarding the necessary testing of all the lifting equipment in his charge.

This Chart should include the following information:-

1. The Manufacturer of the Lifting Equipment.

2. S.P.I.N. Number.

3. Location of the Lifting Equipment.

4. Statutory Inspection Period.

5. Name and Numbers of Applicable Lifting Tackle.

6. Schedule and History Card Nos.

In most cases the necessary testing of lifting equipment can be implemented by specialized companies, or by the appropriate insurance company.

In the Third World it is normally the responsibility of the Maintenance Manager or Superintendent to arrange all testing in-house by means of specific weights, ie: concrete blocks etc., as there are no Statutory laws that apply. Normally the Maintenance Manager is responsible for any injury or fatality occuring as a consequence of untested lifting equipment.

10TH JUNE 1986

WATER & POWER STATION

RUMAIS

CRANE & LIFTING EQUIPMENT TEST PROGRAMME

COMPANY: "EMIRATE SAFETY SERVICES"

MAINTENANCE ACTIVITY SCHEDULED BY THE YEAR

YEAR: 19

ITEM	COMPANY	SHORT DESCRIPTION OF ACTIVITY	REMARKS
1	O-UU-M 01	MAIN HOIST TURBINE HALL	SWL/100/TL125 TONS
2		AUXILIARY HOIST	SWL/20/TL25 TONS
3	O-UU-M 21	CHLORINAT.STATION CRANE	
4		ELECTRIC HOIST	SWL/3/TL3.75 TONS
5	O-UU-M 23	W.T.P. CRANE	
6			SWL/1/TL/1.25 TONS
7	O-UU-M 01	STEAM GENERATION PLANT LIFT	
8			SWL/2/TL2.5 TONS
9	O-UV-M 02	W.T.P. HYDRAULIC LIFT	
10			SWL/3/TL3.75 TONS
11		P&H OMEGA	
12		MOBILE CRANE	SWL/30/TL37.5 TONS
13	O-UU-P 04	WORKSHOP O.H. TRAVEL	
14		CRANE – TRAVEL CRAB	SWL/15/TL18.75 TONS
15	O-UU-M 02	SCREENING PLANT GANTRY CRANE	SWL/15/TL18.75 TONS
16			SWL/5/TL6.25 TONS
17		CHAIN HOIST	
18		(3 NOS.)	SWL/3/TL3.75 TONS
19		CHAIN HOIST	
20			SWL/10/TL12.5 TONS
21		CHAIN BLOCK	
22		(2 NOS.)	SWL/0.75/TL0.95 TONS
23		CHAIN BLOCK	
24		(2 NOS.)	SWL/1.5/TL1.87 TONS
25		CHAIN RATCHET	
26		(1 NO.)	SWL/1.5/TL1.87 TONS
27		STEEL WIRE	
28		ROPE SLING	SWL/1/4 – 2½/TL3 TONS
29		STEEL WIRE	
30		ROPE SLING	SWL/5-20/TL25 TONS
31		NYLON SLING	
32		(4 NOS.)	SWL/1/4 – 6/TL7.5 TONS
33		SHAKLE FORGED STEEL	
34		1/4"-2" (8 NOS.)	
35		FORK LIFT	
36			SWL/5.0/TL6.25 TONS
37		FORK LIFT	
38			

WEEK: 1 2 3 4 5 6 7 8 9 10 11 12 13 14 15 16 17 18 19 20 21 22 23 24 25 26 27 28 29 30 31 32 33 34 35 36 37 38 39 40 41 42 43 44 45 46 47 48 49 50 51 52

Months: JAN FEB MAR APR MAY JUN JUL AUG SEPT OCT NOV DEC

SCHEDULED
SCHEDULE COMPLETED

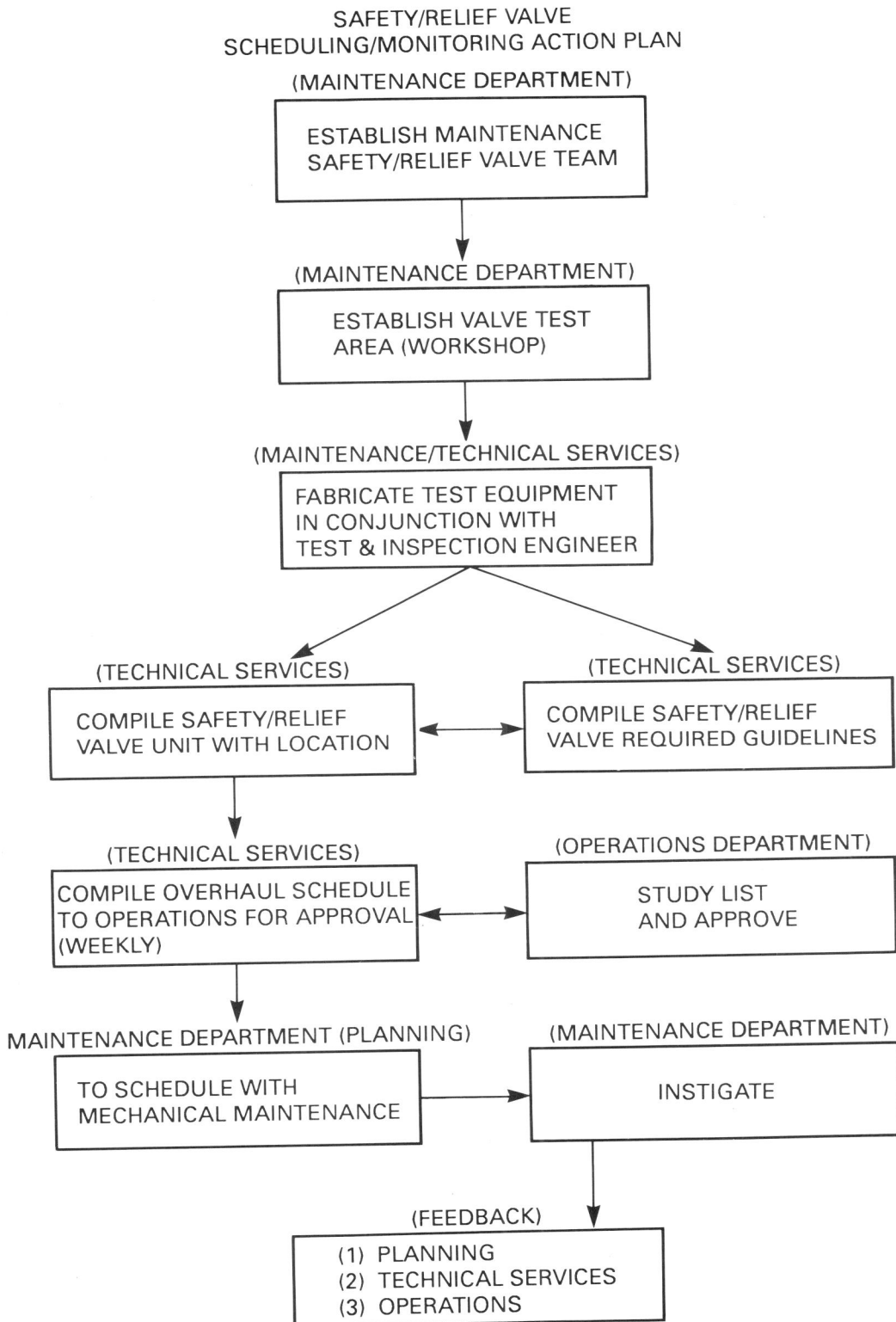

SAFETY/RELIEF VALVE
SCHEDULING/MONITORING ACTION PLAN

(MAINTENANCE DEPARTMENT)

```
┌─────────────────────────────┐
│  ESTABLISH MAINTENANCE      │
│  SAFETY/RELIEF VALVE TEAM   │
└─────────────────────────────┘
```

(MAINTENANCE DEPARTMENT)

```
┌─────────────────────────────┐
│  ESTABLISH VALVE TEST       │
│  AREA (WORKSHOP)            │
└─────────────────────────────┘
```

(MAINTENANCE/TECHNICAL SERVICES)

```
┌─────────────────────────────┐
│  FABRICATE TEST EQUIPMENT   │
│  IN CONJUNCTION WITH        │
│  TEST & INSPECTION ENGINEER │
└─────────────────────────────┘
```

(TECHNICAL SERVICES)   (TECHNICAL SERVICES)

```
┌─────────────────────────┐      ┌─────────────────────────┐
│  COMPILE SAFETY/RELIEF   │ ←──→ │  COMPILE SAFETY/RELIEF   │
│  VALVE UNIT WITH LOCATION│      │  VALVE REQUIRED GUIDELINES│
└─────────────────────────┘      └─────────────────────────┘
```

(TECHNICAL SERVICES)   (OPERATIONS DEPARTMENT)

```
┌─────────────────────────────┐   ┌─────────────────────────┐
│  COMPILE OVERHAUL SCHEDULE  │←─→│  STUDY LIST              │
│  TO OPERATIONS FOR APPROVAL │   │  AND APPROVE            │
│  (WEEKLY)                   │   └─────────────────────────┘
└─────────────────────────────┘
```

MAINTENANCE DEPARTMENT (PLANNING)   (MAINTENANCE DEPARTMENT)

```
┌─────────────────────────────┐      ┌─────────────────────────┐
│  TO SCHEDULE WITH           │ ───→ │  INSTIGATE              │
│  MECHANICAL MAINTENANCE     │      │                         │
└─────────────────────────────┘      └─────────────────────────┘
```

(FEEDBACK)

```
┌─────────────────────────────┐
│  (1)  PLANNING              │
│  (2)  TECHNICAL SERVICES    │
│  (3)  OPERATIONS            │
└─────────────────────────────┘
```

WORK ORDER INPUT, OUTPUT, FLOW DIAGRAM

The Maintenance Work Input and Output flow diagram indicates that for given information, ie: Work Orders with the attached informa- tion instruction.

    1. MANHOURS PER ENGINE OR UNIT EQUIPMENT.

    2. SPARES UTILIZED.

    3. DATE & TIME WORK ORDERS ISSUED AND COMPLETED.

    4. JOB DESCRIPTION.

The following critical information including accumulative factors are available:-

A   1. MANHOUR COSTS PER HOUR INCLUDING OVERTIME.

    2. ACCUMULATIVE MANHOUR COSTS PER ENGINE OR UNIT.

    3. INDIVIDUAL COSTS.

B   1. SPARE OR REPLACEMENT COSTS INDIVIDUAL.

    2. ACCUMULATIVE SPARE USED AND ACCUMULATIVE COSTS.

    3. STOCK NUMBERS AND MANUFACTURER'S NAME LOCATION.

    4. RE-ORDER CAPABILITY. LEAD OR LAG TIME CONSIDERATION.

    5. INFLATION COST PERCENTAGE RATIO.

C   1. AVAILABILITY FACTORS.

    2. SYSTEM OR ITEM PLANT EFFECTIVENESS.

    3. RELIABILITY CALCULATIONS FACTOR.

    4. MEAN TIME BEFORE FAILURE

D   1.   MONITOR JOB FUNCTION.

         2.   MONITOR ENGINE OR UNIT BREAKDOWN PEAKS.

         3.   UPDATE PLANNED PREVENTATIVE MAINTENANCE
             SYSTEM TO ELIMINATE PEAKS/PEAK LOPPING.

ACCUMULATIVE FACTORS

         A.   SPARES.

         B.   LABOUR.

         C.   STATISTICAL LIFE CURVE COSTS.

FISCAL BUDGET (%)

         A.   LABOUR.

         B.   SPARES.

         C.   TOTAL COST.

If necessary, a figure for the total cost applicable to the non-
availability of a production unit can be obtained if the cost per
unit per Production hour is known.

WORK ORDER INPUT AND OUTPUT FLOW DIAGRAM

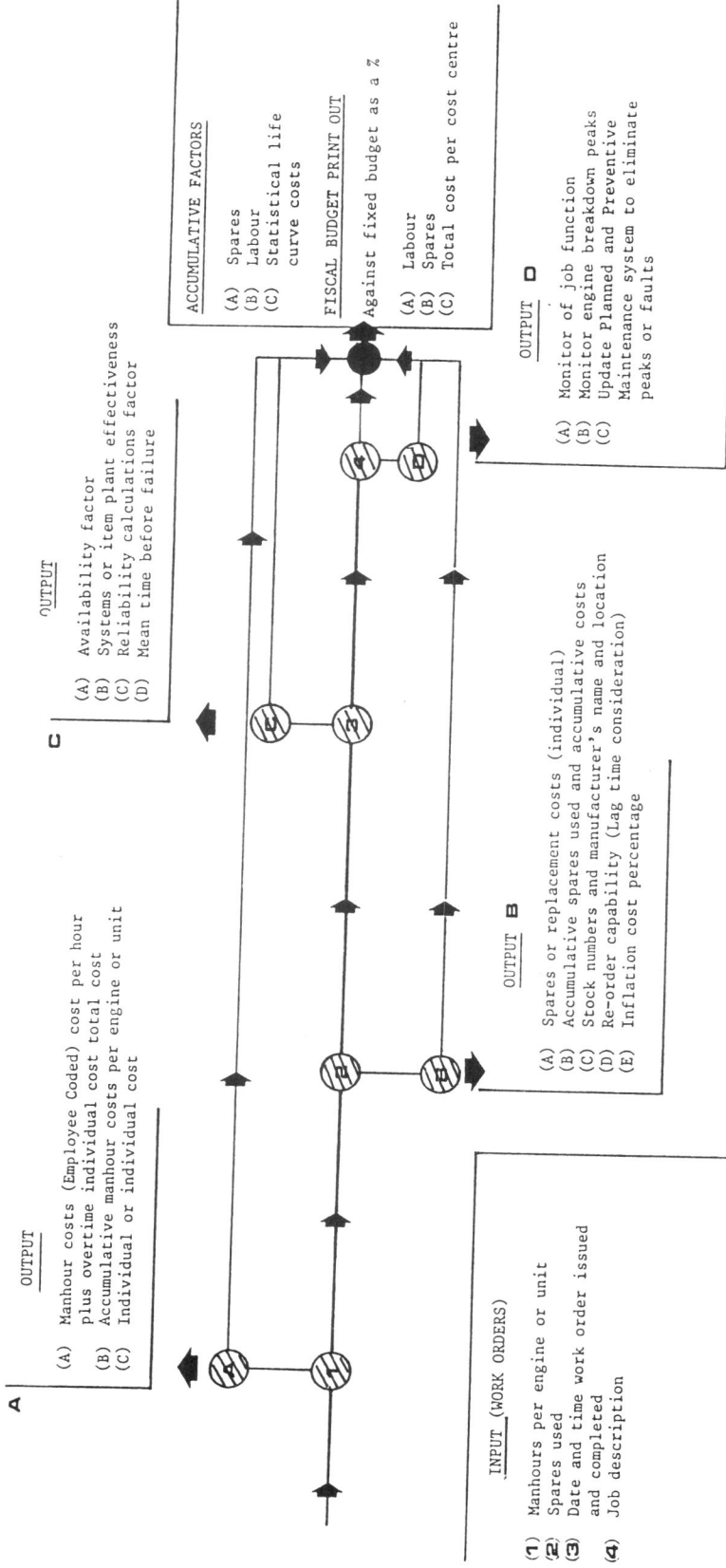

A

OUTPUT

(A) Manhour costs (Employee Coded) cost per hour
    plus overtime individual cost total cost
(B) Accumulative manhour costs per engine or unit
(C) Individual or individual cost

C

OUTPUT

(A) Availability factor
(B) Systems or item plant effectiveness
(C) Reliability calculations factor
(D) Mean time before failure

ACCUMULATIVE FACTORS

(A) Spares
(B) Labour
(C) Statistical life
    curve costs

FISCAL BUDGET PRINT OUT

Against fixed budget as a %

(A) Labour
(B) Spares
(C) Total cost per cost centre

OUTPUT **D**

(A) Monitor of job function
(B) Monitor engine breakdown peaks
(C) Update Planned and Preventive
    Maintenance system to eliminate
    peaks or faults

OUTPUT **B**

(A) Spares or replacement costs (individual)
(B) Accumulative spares used and accumulative costs
(C) Stock numbers and manufacturer's name and location
(D) Re-order capability (Lag time consideration)
(E) Inflation cost percentage

INPUT (WORK ORDERS)

(1) Manhours per engine or unit
(2) Spares used
(3) Date and time work order issued
    and completed
(4) Job description

AIR CONDITIONING WINDOW SPLIT UNITS CHECK LIST

MANUFACTURER	UNIT TYPE	SERIAL NO.
LOCATION	AREA	ROOM NO.
I.D. NO.	DATE INSTALLED	
DATE OF LAST MAINTENANCE		
NORMAL UNIT OPERATION	VOLTS	AMPS

MONTHLY PREVENTIVE MAINTENANCE

ITEM A		DATE	REMARKS
1	CHECK, CLEAN AND WASH, RETURN AIR FILTER		
2	CHECK AND CLEAN UNIT DRAIN SYSTEM		
3	CHECK FOR WATER LEAKS		
4	CHECK FOR ABNORMAL VIBRATION		
5	CHECK OPERATION OF COMPRESSOR OVERLOADED RELAY		
6	CHECK "OPERATION" SWITCH		
7	CHECK OPERATION OF TEMPERATURE THERMOSTAT		
8	CHECK FOR AIR CONDITIONER GROUND WIRE		
9	CHECK UNIT FOR CORRECT FUSE		
10	CHECK RUNNING AMPS = AND RUNNING VOLTS =		
11	CHECK OPERATION UNIT		
12	CHECK FOR REFRIGERANT LEAKS		

SIX MONTHS PLANNED MAINTENANCE OVERHAUL

ITEM B		DATE	REMARKS
1	CHECK UNIT FOR OPERATION. NOTE ANY DEFECTS		
2	REMOVE UNIT TO WORKSHOP AND INSTALL STAND-BY UNIT		
3	CLEAN UNIT		
4	CLEAN AND INSPECT EVAPORATOR AND CONDENSER UNITS		
5	CLEAN AND INSPECT MOTOR FAN AND SIROCCO FAN		
6	CHECK FOR CRACKS AND BLADES RUBBING CHECK OPERATION OF FAN MOTOR, AND HOLDING DOWN BOLTS FOR TIGHTNESS		
7	GREASE AND LUBRICATE ALL BEARINGS		
8	CHECK AND INSPECT BEARINGS FOR WEAR		
9	CLEAN AND INSPECT UNIT COMPRESSOR PUMP UNIT DOWN IF REQUIRED		
10	CHECK AND INSPECT FOR REFRIGERANT LEAKS		
11	CHARGE UNIT WITH REFRIGERANT AS NECESSARY		
12	CHECK AND INSPECT ALL ELECTRICAL CONNECTIONS LUBRICATE WITH WD 40		
13	CHECK AND CLEAN AIR FILTER		
14	CHECK AND CLEAN DRAIN SYSTEM		
15	TEST RUN UNIT IN WORKSHOP		
16	CHECK FOR ABNORMAL VIBRATION		
17	CHECK UNIT FUNCTIONS		
18	RE-INSTALL UNIT IN ORIGINAL LOCATION AND RETURN STAND-BY UNIT TO WORKSHOP		
19	CHECK OPERATION OF COMPRESSOR OVERLOAD RELAY		
20	CHECK "OPERATION" SWITCH		
21	CHECK OPERATION OF TEMPERATURE THERMOSTAT		
22	CHECK UNIT FOR GROUND WIRE CHECK UNIT FOR CORRECT FUSE		
23	CHECK UNIT FOR CORRECT ROTATION OF EVAPORATOR AND CONDENSER FAN		
24	CHECK RUNNING AMPS AND VOLTS		
25	CHECK AIR TEMPERATURE BEFORE AND AFTER EVAPORATION		

CONTRACTOR   TECHNICIAN'S NAME

REMARKS   SUPERVISOR

COMPANY'S SUPERVISOR

REMARKS

MODEL_____NO._____COMPRESSOR_____NO._____

CUSTOMER'S NAME AND ADDRESS_____DATE:_____

1. Is the rotation direction of the evaporator fan correct?   ☐

2. Is the rotation direction of the condenser fan correct?   ☐

3. Are there any abnormal compressor sounds?   ☐

4. Has the unit been operated for at least twenty (20) minutes?   ☐

5. Check Room Temperature:

   Inlet: DB_____DC,   WB_____DC
   Outlet: DB_____DC,   WB_____DC

6. Check Ambient Temperature:

   Inlet: DB_____DC,   WB_____DC
   Outlet: DB_____DC,   WB_____DC

7. Check Suction Line Temperature and Superheat:

   Suction Line Temperature:_____DC
   Superheat:_____DC

8. Check Pressure:

   Discharge Pressure:_____kg/cm²
   Suction Pressure:_____kg/cm²
   Oil Pressure:_____kg/cm²

9. Check Voltage:

   Rating Voltage:_____V,
   Operation Voltage: R·S_____V,   R·T_____V,   S·T_____V,
   Starting Voltage:_____V,
   Phase Imbalance: 1 V/V$_m$ =_____

10. Check Compressor Input and Running Current:

   Input: _____kw
   Running Current:_____A

11. Is the refrigerant charge adequate?   ☐

12. Do the operation control devices operate correctly?   ☐

13. Do the safety devices operate correctly?   ☐

14. Has the unit been checked for refrigerant leakage?   ☐

15. Is the unit clean inside and outside?   ☐

16. Are all cabinet panels fixed?   ☐

17. Are all cabinet panels free from rattling?   ☐

18. Is the filter clean?   ☐

19. Is the condenser clean?   ☐

20. Are the stop valves open?   ☐

MAINTENANCE SCHEDULE

Operation	1 Week	3 Months	1 Year	5 Years	Notes
Compressor					
Check Oil Level	X				
Charge Oil			X		
Charge Oil Filter		X			
Check Shaft Seal Temp	X				
Check Seal Leakage		X			
Check Crankcase Heater			X		
Overhaul Compressor				X	
Drive Motors					
Check for Grease Leakage	X				
Replace Greaser			X		
Overhaul				X	
Couplings					
Check for Wear & Damage			X		
Condenser					
Motor Check Grease Leakage	X				
Replace Motor Grease			X		
Check Fan Blade Pitch		X			
Check Fan Bearings	X				
Check Air Path Cleanliness	X				
Check Belts		X			

MAINTENANCE SCHEDULE (cont'd)

Operation	1 Week	3 Months	1 Year	5 Years	Notes
Chiller					
Internal Check & Clean				X	
Liquid Suction H.E.					
Internal Check & Clean				X	
Oil Separator					
Check Float Valve			X		
Internal Inspect				X	
Oil Cooler					
Inspection			X		
Filter Drier					
Check Moisture Level	X				
Replace Filter Elements			X		
Valves					
Inspection			X		
Replace Overhaul				X	
Instruments					
Inspection Check			X		
Replace-Overhaul				X	
Check Refrigerant Level	X				

PETROLEUM DEVELOPMENT (OMAN) LLC

MAINTENANCE PLANNED MONTHLY SCHEDULE

PAGE [ ] MONTH [ ] 198[ ]

A EQUIPMENT IDENTIFICATION NUMBER	B ELECTRICAL UNIT NAME	C PLANNED OR UN-PLANNED MAINTENANCE	D DATE OF SCHEDULED MAINTENANCE	E ACTUAL DATE OF MAINTENANCE	F MAN HOURS AS APPLICABLE	G JOB NO	H WORK CARRY OVER	I DOWNTIME AND COMMENTS P.T.O.	COLOUR CODE
1									
2									
3									
4									
5									
6									
7									
8									
9									
10									
11									
12									
13									
14									
15									
16									
17									
18									
19									
20									
21									
22									
23									
24									
25									
26									
27									
28									
30									
31									
32									
33									
34									
35									
36									
37									
38									
39									
40									
41									
42									
43									
44									

PLANNED/PREVENTATIVE MAINTENANCE FLOW DIAGRAM

The Planned, Preventative Maintenance Flow Diagram is an extremely useful document for the creation of a Maintenance System or the expansion of an existing one.

The person or persons assigned to either the creation or expansion of a Maintenance System would commence with the following requirements or recommendations as per the applicable headings:-

1. TECHNICAL LIBRARY

2. EQUIPMENT RECORD DATA CARDS

3. PLANT MAINTENANCE SCHEDULE 403

4. MAINTENANCE MONTHLY SCHEDULE RECORDING

5. ISSUE FORM I.M.F.

6. PLANNED MAINTENANCE TASK SHEET 2MDI

7. PREVENTATIVE MAINTENANCE TASK SHEET 5M2

8. EQUIPMENT FAILURE SHEET 6M4

9. MAINTENANCE TROUBLE SHOOTING GUIDE

The remaining information applicable to the Planned Preventative Maintenance Flow Sheet is returning information, ie: feedback, once the issue documentation and base information has been compiled.

The returning information, or feedback, would be utilized to supplement the issue and base documentation for system improvement and necessary overall monitoring.

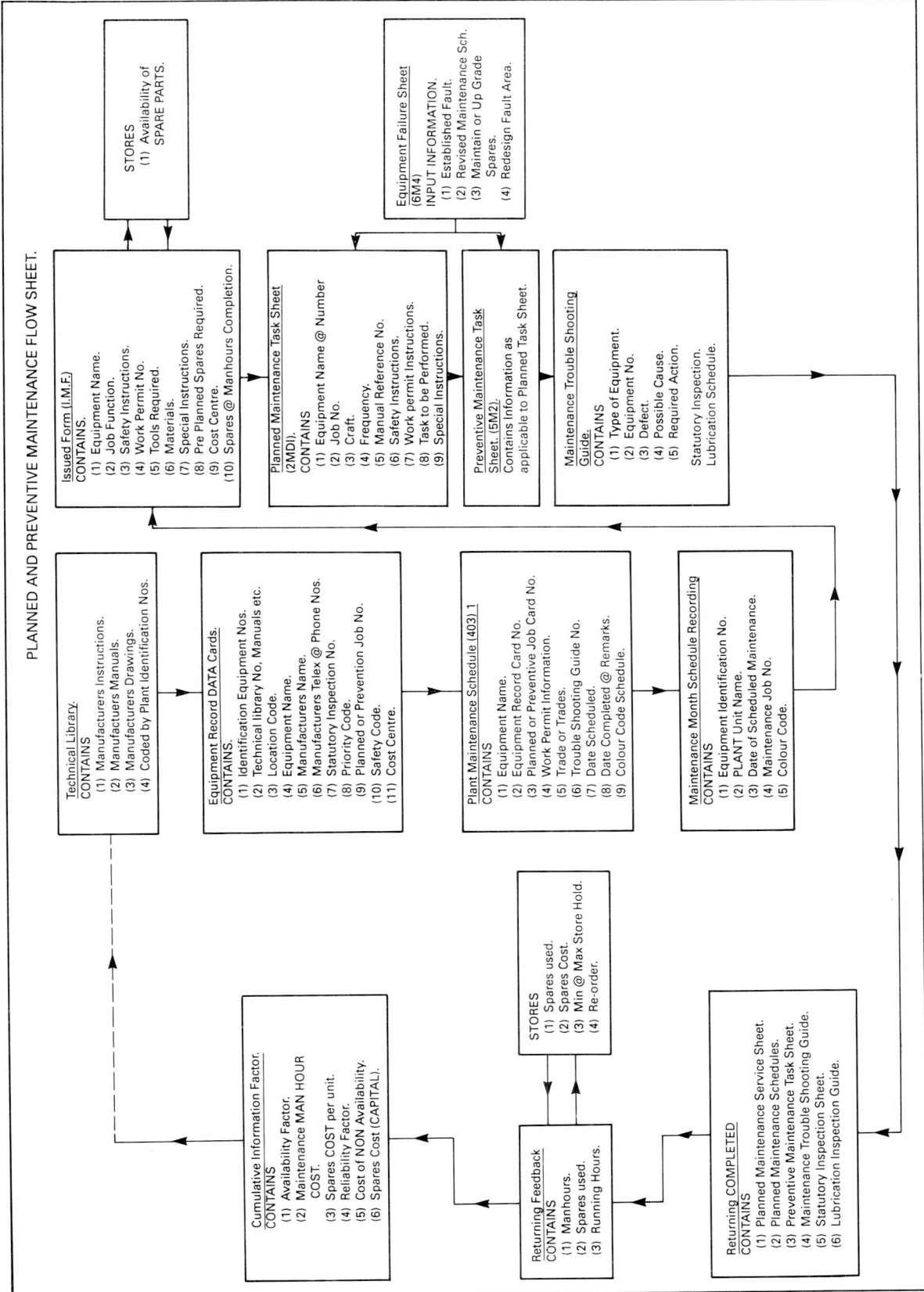

PLANNED AND PREVENTIVE MAINTENANCE FLOW SHEET.

STORES
(1) Availability of SPARE PARTS.

Equipment Failure Sheet (6M4)
INPUT INFORMATION.
(1) Established Fault.
(2) Revised Maintenance Sch.
(3) Maintain or Up Grade Spares.
(4) Redesign Fault Area.

Issued Form (I.M.F.)
CONTAINS.
(1) Equipment Name.
(2) Job Function.
(3) Safety Instructions.
(4) Work Permit No.
(5) Tools Required.
(6) Materials.
(7) Special Instructions.
(8) Pre Planned Spares Required.
(9) Cost Centre.
(10) Spares @ Manhours Completion.

Planned Maintenance Task Sheet (2MDI).
CONTAINS
(1) Equipment Name @ Number
(2) Job No.
(3) Craft.
(4) Frequency.
(5) Manual Reference No.
(6) Safety Instructions.
(7) Work permit Instructions.
(8) Task to be Performed.
(9) Special Instructions.

Preventive Maintenance Task Sheet. (5M2).
Contains Information as applicable to Planned Task Sheet.

Maintenance Trouble Shooting Guide.
CONTAINS
(1) Type of Equipment.
(2) Equipment No.
(3) Defect.
(4) Possible Cause.
(5) Required Action.

Statutory Inspection.
Lubrication Schedule.

Technical Library.
CONTAINS
(1) Manufacturers Instructions.
(2) Manufacturers Manuals.
(3) Manufacturers Drawings.
(4) Coded by Plant Identification Nos.

Equipment Record DATA Cards.
CONTAINS.
(1) Identification Equipment Nos.
(2) Technical library No, Manuals etc.
(3) Location Code.
(4) Equipment Name.
(5) Manufacturers Name.
(6) Manufacturers Telex @ Phone Nos.
(7) Statutory Inspection No.
(8) Priority Code.
(9) Planned or Prevention Job No.
(10) Safety Code.
(11) Cost Centre.

Plant Maintenance Schedule (403) 1
CONTAINS
(1) Equipment Name.
(2) Equipment Record Card No.
(3) Planned or Preventive Job Card No.
(4) Work Permit Information.
(5) Trade or Trades.
(6) Trouble Shooting Guide No.
(7) Date Scheduled.
(8) Date Completed @ Remarks.
(9) Colour Code Schedule.

Maintenance Month Schedule Recording
CONTAINS
(1) Equipment Identification No.
(2) PLANT Unit Name.
(3) Date of Scheduled Maintenance.
(4) Maintenance Job No.
(5) Colour Code.

Cumulative Information Factor.
CONTAINS
(1) Availability Factor.
(2) Maintenance MAN HOUR COST.
(3) Spares COST per unit.
(4) Reliability Factor.
(5) Cost of NON Availability.
(6) Spares Cost (CAPITAL).

STORES
(1) Spares used.
(2) Spares Cost.
(3) Min @ Max Store Hold.
(4) Re-order.

Returning Feedback
CONTAINS
(1) Manhours.
(2) Spares used.
(3) Running Hours.

Returning COMPLETED
CONTAINS
(1) Planned Maintenance Service Sheet.
(2) Planned Maintenance Schedules.
(3) Preventive Maintenance Task Sheet.
(4) Maintenance Trouble Shooting Guide.
(5) Statutory Inspection Sheet.
(6) Lubrication Inspection Guide.

EXTERNAL MAINTENANCE PROGRAMME

As Maintenance Managers are aware it is not always cost effective to have all skills available in-house, ie; Lift Maintenance, calibration of specialized equipment etc.

If outside contractors or agents are utilized it will be necessary to control and supervise their activity so as to ensure they conform to the agreed specification of the maintenance contract.

An External Maintenance Programme Chart is required which has to be scheduled by the year.
This chart should also include the following:-

1. Applicable Contract Maintenance Company.

2. Equipment Name.

3. Description of Maintenance Company Activity.

4. Scheduled Maintenance Period / Week or Month.

5. Remarks applicable to the Maintenance instig-
    ated. Parts Change out etc.

It will also be necessary to enter all information applicable to the Maintenance Service or overhaul on the In-House History card or document including Man-hours and spare part costs etc.

EXTERNAL MAINTENANCE PROGRAMME

YEAR: 1986

MAINTENANCE ACTIVITY SCHEDULED BY THE YEAR

Weeks 1–52 across JAN · FEB · MAR · APR · MAY · JUN · JUL · AUG · SEPT · OCT · NOV · DEC

ITEM	COMPANY	SHORT DESCRIPTION OF ACTIVITY	REMARKS
		MACHINE ROOM	
1		PS OF EMERG. CALL & LTG.	INSPECT
2		LEAKAGE PUMP	INSPECT (FUNCTIONAL TEST)
3		PRESSURE PIPELINE	INSPECT FOR LEAKS
4		EL. IND. LAMPS, PUSH BUTTON	INSPECT
5		GEAR & MOTOR	INSPECT OIL LEVEL
6		BRK. PULLY & GEAR HOUSING	CLEAN
7		DRIVE & DEFLECTOR	INSPECT WEAR
8		BRAKE	INSPECT & ADJUST
9		BELTS	INSPECT & ADJUST
10		SPEED CONTROLLER	CLEAN INTERIOR
11		CONTROLLER & C BRUSHES	INSPECT
12		EMERGENCY LIMIT SWITCH	INSPECT
13	BENHARB	FLOOR SELECTOR CONTACTS	INSPECT
14	CORPORAT.	HOISTING & GOVERNOR CBL	INSPECT
15	(BOILER	ALL CONTACTORS & RELAYS	INSPECT ADJUSTMENT
16	& W.T.	ALL TIMERS	CLEAN
17	LIFT)	CONTROLLER	TIGHT
18		MOUNTG & CONNCTG SCREWS	INSPECT IF DAMAGED
19		TRAVELLING CABLE	INSPECT & SERVICE
20		HOISTWAY SWITCHES	INSPECT & SERVICE
21		GUIDERAIL FASTG. & GAUGES	CHANGE OIL
22		GEAR UNIT & MOTOR	CLEAN INTERIOR
23		MOTOR & GENERATOR	INSPECT
24		COLLECTOR & C BRUSHES	INSPECT & LUBRICATE
25		GOVERNOR	INSPECT
26		OP. CONTROL/MOTOR OIL	INSPECT UNIFORM TENSION
27		HOISTING	
		LANDING DOOR	
28		LOCKS/CONTACTS	INSPECT
29		MOVING PARTS	LUBRICATE
30		DOOR SILLS & GUIDE SHOES	CLEAN
		COUNTER WEIGHT	
31		GUIDE RAIL OILERS	TOP UP
32		FASTING OF FLOOR SELECTOR	INSPECT
33		GUIDE SHOES/ROLLER GUIDE	INSPECT

Legend:
SCHEDULED (black) ◣
SCHEDULE COMPLETED (open box) ☐

MAIN MAINTENANCE ACTIVITIES. OUTAGE SCHEDULE.

In all major Installations it is necessary to plan all
Maintenance Activities applicable to a MAJOR OUTAGE PROGRAMME
which normally occurs once a year.

These Planned Maintenance Outages are necessary for Statutory
Inspections, ie: Boiler and Pressure Vessel inspection tests etc.
The Outages also include necessary rectifications and
modifications to plant and equipment which were not possible
during the year due to production requirements.

In order to obtain the least production disruption and without
infringement on the statutory requirements the OUTAGE SCHEDULE
document should be circulated throughout Senior Management, ie:
Plant Manager, Production Manager etc. for their comments and
requirements.

It is possible to reschedule the Outage Programme but this is
subject to statutory requirements.

WATER & POWER STATION
RUWAIS

MAIN MAINTENANCE ACTIVITIES APPLICABLE

TO THE PERIOD 1986/1987

1 9 8 6

#	MAINTENANCE ACTIVITY	Schedule (JAN–DEC)	REMARKS
1	NO.1 BOILER MAINTENANCE/OVERHAUL	JAN	ANNUAL MAINTENANCE/OVERHAUL
2	NO.1 STEAM TURBINE MAINTENANCE OVERHAUL	JAN–FEB	ANNUAL MAINTENANCE/OVERHAUL
3	NO.1 DESAL. MAINTENANCE OVERHAUL	FEB–MAR	ANNUAL MAINTENANCE/OVERHAUL
4	G.E. GAS TURBINE MAINTENANCE OVERHAUL TP01	APR	MAJOR MAINTENANCE/OVERHAUL
5	FERTIL PUMPS/PLANT MAINTENANCE OVERHAUL		ANNUAL MAINTENANCE
6			
7			
8			
9			
10			
11			
12			
13	PREVENTIVE MAINTENANCE NO.2 BOILER/TURBINE	AUG	TURBINE CONDENSOR FEED PUMPS
14	PREVENTIVE MAINTENANCE NO.1 BOILER/TURBINE	AUG	TURBINE CONDENSOR FEED PUMPS ETC.
15	NO.2 BOILER MAINTENANCE/OVERHAUL	OCT	OUTAGE AS PER HYDROCRACKER
16	NO.2 STEAM TURBINE MAINTENANCE OVERHAUL	OCT–NOV	OUTAGE AS PER HYDROCRACKER
17	NO.1 BOILER MAINTENANCE/OVERHAUL	NOV	OUTAGE AS PER HYDROCRACKER
18	NO.1 STEAM TURBINE MAINTENANCE OVERHAUL	NOV–DEC	OUTAGE AS PER HYDROCRACKER
19			

1 9 8 7

#	MAINTENANCE ACTIVITY	Schedule (JAN–DEC)	REMARKS
1	NO.1 DESAL MAINTENANCE/OVERHAUL	JAN	ANNUAL MAINTENANCE/OVERHAUL
2	NO.2 DESAL MAINTENANCE/OVERHAUL	FEB	ANNUAL MAINTENANCE/OVERHAUL
3			
4			
5	FERTIL PUMPS PLANT MAINTENANCE/OVERHAUL	APR	ANNUAL MAINTENANCE
6			
7			
8			
9			
10			
11			
12			
13			
14			
15	PREVENTIVE MAINTENANCE NO.2 BOILER/TURBINE	AUG	
16	PREVENTIVE MAINTENANCE NO.1 BOILER/TURBINE	AUG	
17		OCT	OUTAGE AS PER HYDROSKIMMER
18	NO.2 BOILER MAINTENANCE/OVERHAUL	OCT	OUTAGE AS PER HYDROSKIMMER
19	NO.2 STEAM TURBINE	NOV	OUTAGE AS PER HYDROSKIMMER
20			OUTAGE AS PER HYDROSKIMMER
21			

6 Maintenance vibration analysis

VIBRATION ANALYSIS

GENERAL

In conjunction with an established Planned and Preventative Maintenance System, Vibration Analysis, conducted on a regular and professional level, can be an extremely useful aid in Plant and Equipment Monitoring.

It will enable the Maintenance Engineer or Maintenance Technician to detect at an early stage any abnormal Plant and Equipment operating parameters if the Vibration readings are interpreted correctly.

As most Maintenance Engineers and Technicians are aware, there are two Vibrational Systems:-

    (a)   Installed Supervisory Vibration monitoring.

    (b)   Portable Vibration Analysis.

Installed Supervisory Vibration Monitoring is normally installed in large mass rotating equipment such as Steam and Gas Turbines, large pumps and compressors etc.

This type of Vibration monitoring usually incorporates the following:-

    (a)   Alarm Condition.

    (b)   High Alarm Condition.

    (c)   Equipment Trip Condition.

This type of installation is normally subject to Operating Log Sheet scrutinization for any abnormal operating conditions and the necessary Annual Vibration Equipment Calibration.

PORTABLE VIBRATION ANALYSIS

IMPLEMENTATION

Engineers, as with all other professions, are governed by cost
factors. It will normally be necessary for the concerned
Maintenance Manager to justify to his Senior Management why they
should introduce or implement a Portable Vibration Analysis
System with all the additional costs it would involve, ie: the
purchase of the Vibration equipment with associated probes,
training of in-house Maintenance staff, and the necessary
monitoring documentation.

It has been the experience of the writer that the justification
to purchase and implement a portable vibration system can be made
on Plant Availability and Reliability factors involving "Non-
Availability" records with associated cost factors.

EXAMPLE

Availability of a major piece of Production plant with associated
accumulative production factors = 73%. This represents as a
singular factor of 45.36 "hours down" time based on 24 hours
continuous Production. If however we consider the accumulative
nonproduction factors, say of 10 other machines not producing due
to the non-availability of one machine, the end result will be
equivalent to 45.36 x 10 = 453.6 hours of lost production. If
the end cost factor loss was equivalent to 1000 pounds/dollars,
the total loss to the company would equal 453,600. This figure
does not include the loss of productivity of the production staff
while the said production machines are not available.

When approval has been obtained the following main points must be
considered:-

1. What portable Vibration Equipment to purchase?
   What other features should be considered, ie:
   should the unit have its own print-out included
   or is the unit capable of PC connection?
   What type of demonstration will the Agent offer?

2. Training of in-house Maintenance Personnel applicable to the Vibration Equipment which has been purchased. Cost of such training.

3. The selection of the Plant and equipment which will be subject to vibrational regular monitoring. Selection of vibration monitoring areas applicable to each machine, ie: in the vicinity of the holding down bolts etc.

4. Portable Vibration monitoring documentation which should include the following:-

   (a) Equipment Name.

   (b) Area Location.

   (c) S.P.I.N. Number.

   (d) Maintenance Schedule No. including History.

   (e) Priority Scheduling.

   (f) Vibration Equipment Operator.

   (g) Vibration Monitoring Scheduled over a 52 week period.

On completion of the above mentioned, it will then be necessary to undertake Base Signature Readings on the selected Plant and Equipment for necessary comparison at a later stage.

138

VIBRATION ANALYSIS
INTRODUCTION ACTION CHART

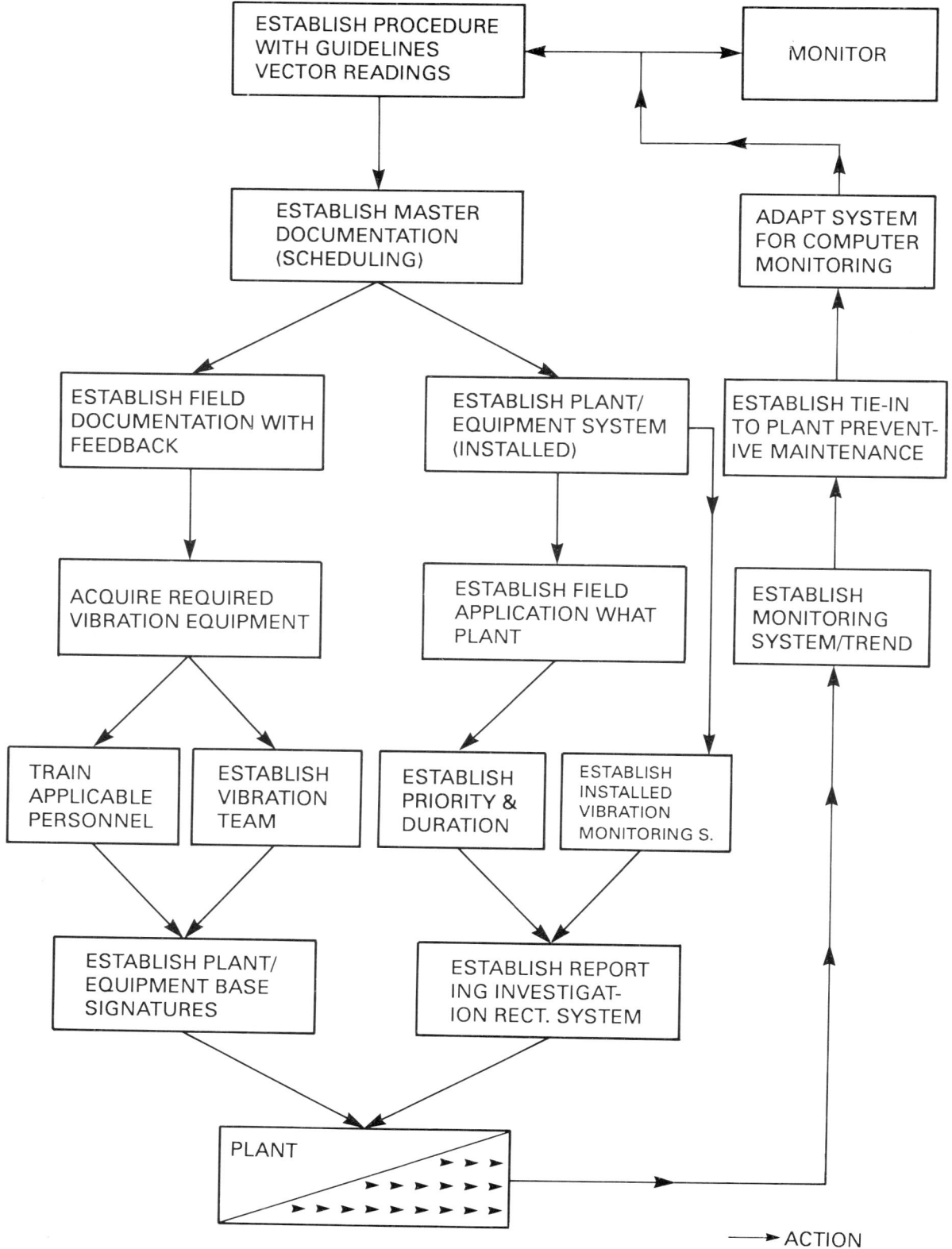

ACTION

VIBRATION ANALYSIS PROGRESS SCHEDULE

ACCUMULATIVE VIBRATION ANALYSIS INVESTIGATION

SHEET NO. 1

PLANT EQUIPMENT

Column headings: ITEM NO. | AREA LOCATION | SCHEDULE NO. | INSTALLED/FITTED | A B C D | (PLANT EQUIPMENT) | WEEK NO. / ITEM NO. / WEEK → 1 2 3 4 5 6 7 8 9 10 11 12 13 14 15 16 17 18 19 20 21 22 23 24 25 26 27 28 29 30 31 32 33 34 35 36 37 38 39 40 41 42 43 44 45 46 47 48 49 50 51 52 | PRIORITY A/B | RENS NO. | REMARKS

Item	Plant Equipment	RENS NO.	Remarks
1	S.W. Circulation pump	1WAG01	MSF I & II
2	S.W. Circulation pump	2WAG01	
3	Sample pump from deaerator	1WCG01	
4	Sample pump from deaerator	2WCG01	
5	Ball pump of heat rejection	1WBG01	
6	Ball pump of heat rejection	2WBG01	
7	Ball pump of heat recov. B.H.	1WBG02	
8	Ball pump of heat recov. B.H.	2WBG02	
9	Ball pump of heat recov. B.H.	1WBG03	
10	Ball pump of heat recov. B.H.	2WBG03	
11	Ball pump of heat recov. B.H.	1WBG04	
12	Ball pump of heat recov. B.H.	2WBG04	
13	Brine recirculation pump	1WNG01A	
14	Brine recirculation pump	2WNG01A	
15	Brine recirculation pump	1WNG01B	
16	Brine recirculation pump	2WNG01B	
17	Brine heater condensate pump	1WQG01A	
18	Brine heater condensate pump	2WQG01A	
19	Brine heater condensate pump	1WQG01B	
20	Brine heater condensate pump	2WQG01B	
21	Blow down pump	1WPG01A	
22	Blow down pump	1WPG01B	
23	Blow down pump	2WPG01B	
24	Blow down pump	1WLG01A	
25	Distillate pump	2WLG01A	
26	Distillate pump	1WLG01B	
27	Distillate pump	2WLG01B	
28	Distillate pump	1WKG0A	
29	High purity water pump	1WKG0B	
30	High purity water pump	2WKG0A	
31	High purity water pump	2WKG0B	
32	High purity water pump	1WHG0A	
33	HCl acid pump	2WHG0A	
34	HCl acid pump	1WHG01B	
35	HCl acid infiction pump	2WHG01B	
36	HCl acid infiction pump	1WHG02A	
37	Antifoam dosing pump	2WHG02A	
38	Antifoam dosing pump	2WHG02B	
39	Antifoam dosing pump	1WHG03A	
40	Sodium sulfite dosing pump	1WHG03B	
41	Sodium sulfite dosing pump	2WHG03A	
42	Sodium sulfite dosing pump		
43	Sodium sulfite dosing pump		

(PLANT EQUIPMENT)

VIBRATION ANALYSIS CONDUCTED AS PER SCHEDULE 52 WEEKS   (A) = CRITICAL PLANT

SCHEDULED VIBRATION (INVESTIGATION/ANALYSIS)

VIBRATION ANALYSIS PROGRESS SCHEDULE

ACCUMULATIVE VIBRATION ANALYSIS INVESTIGATION

PLANT EQUIPMENT

ITEM NO.	AREA LOCATION NO.	SCHEDULE NO.	INSTALLED FITTED	PLANT EQUIPMENT (WEEK NO. 1–52)	PRIORITY A/B	REMARKS
B	C	D		WEEK	A/B	
B1	9	F		FORCED DRAUGHT FAN	A	
B2	7	F		FORCED DRAUGHT FAN	A	
R2	3 1/F			GAS TURBINE 90TP02	B	HIGH SIGNATURE (15)
R1	8 1/F			GAS TURBINE 90TP01	B	
B1	6	F		BOILER FEED PUMP	B	
B1 12		F		BOILER FEED PUMP	B	
B2 11		F		BOILER FEED PUMP	B	
B2 12		F		BOILER FEED PUMP	B	
T1 19	1/F			STEAM TURBINE	A	
T2 20	1/F			STEAM TURBINE	A	

VIBRATION ANALYSIS CONDUCTED AS PER SCHEDULE 52 WEEKS

(A) = CRITICAL PLANT

◤ SCHEDULED VIBRATION (INVESTIGATION/ANALYSIS)

LUBRICATION SCHEDULE

The need for a carefully controlled lubricating system is self evident - Preventive Maintenance.

Failure to lubricate machine parts at the right intervals of time will result in plant breakdowns. Oils and greases are specified by the manufacturer after careful consideration of plant duties and ratings, and the need to use the specified or equivalent lubricants cannot be stressed highly enough. Failures can occur due to overfilling certain types of bearings, particularly the over-greasing of ball and roller bearings, and it is therefore important that oil and grease levels are strictly maintained.

In some companies, lubrication is a function of the Maintenance Department; in others it is the responsibility of the operating staff. Whichever system is in use, a comprehensive schedule is necessary to ensure that the plant is at all times lubricated.

The Preventive Maintenance form must indicate the following:-

(1) Plant item
(2) Plant number
(3) Location
(4) Points of lubrication
(5) Method of application
(6) Recommended lubrication
(7) Frequency of application
(8) Frequency of sampling of oil change

7 Maintenance management

MANAGEMENT BY OBJECTIVES FOR THE YEAR

MAINTENANCE DEPARTMENT

PURPOSE.

To PROVIDE A SAFE, ENHANCED, EFFICACIOUS, MAINTENANCE SERVICE TO OBTAIN OPTIMUM PLANT AVAILABILITY/RELIABILITY FACTORS. THIS SERVICE WILL BE COST EFFECTIVE AND HARMONIOUS.

<u>MANAGEMENT BY OBJECTIVES FOR THE YEAR</u>

<u>MAINTENANCE DEPARTMENT</u>

<u>OBJECTIVE NO.1</u>

MANPOWER COSTS (INCLUDING OVERTIME) TO BE REDUCED BY FULL MAN-POWER OPTIMISATIONS.

OVERTIME TO BE REDUCED TO 2% PER MONTH FROM APRIL 1986.

(A) ANNUAL SAVING (1) EQUATED TO THE
    1985 OVERTIME EXPENDITURE   :   51,819.00 DHS.

(B) SAVING AS EQUATED TO THE 1986
    APPROVED BUDGET   :   128,900.00 DHS.

(C) LABOUR OPTIMISATION OVER THE 1986
    PERIOD (CASCADE PRINCIPLE) SAVING : 1,737,422.00 DHS.
    (APPROX.)

    <u>TOTAL APPROXIMATE SAVING</u> (APPROX.) : 1,780,000.00 DHS.

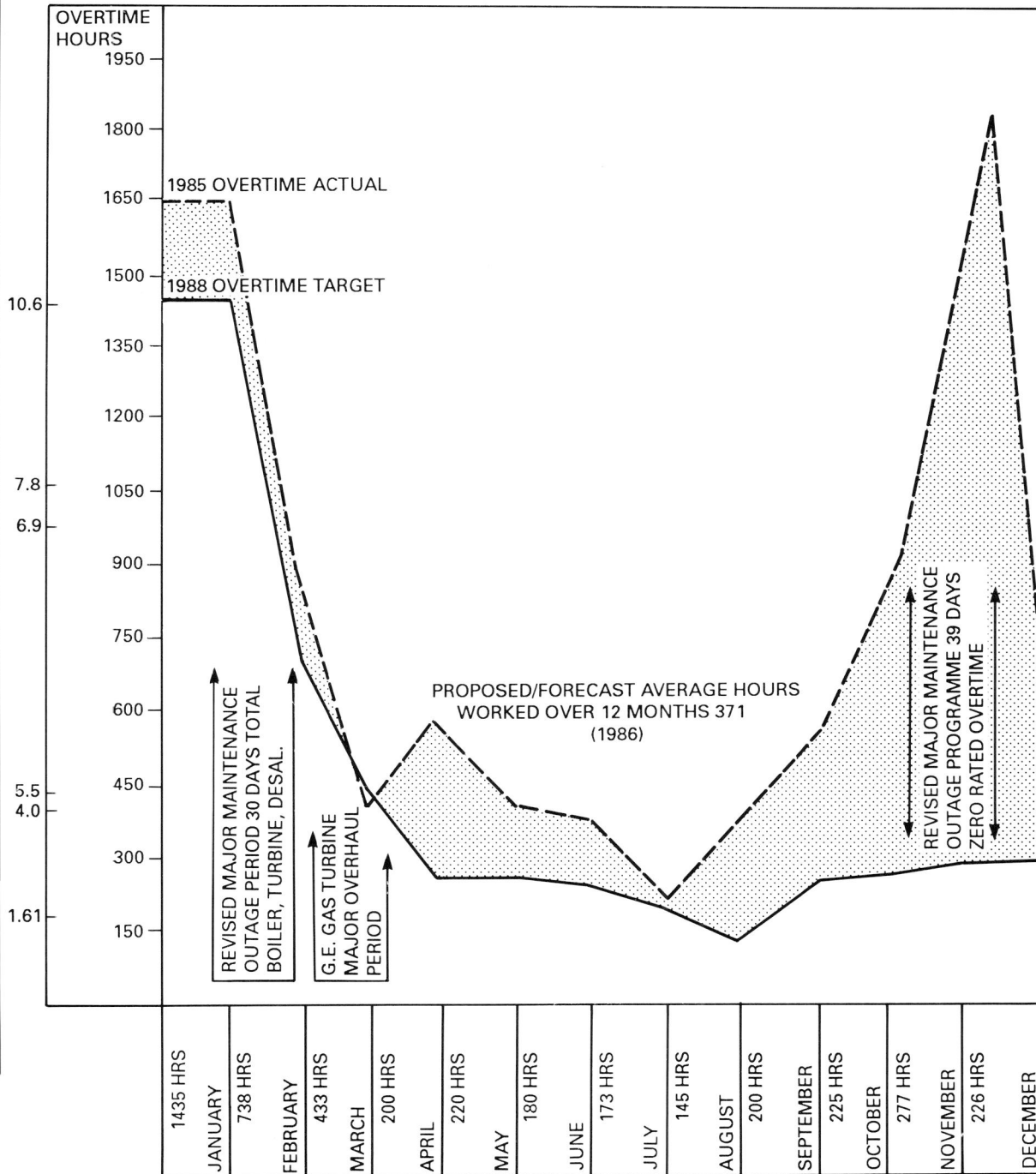

MANAGEMENT BY OBJECTIVES FOR THE YEAR

MAINTENANCE DEPARTMENT

YARDSTICK FOR OBJECTIVE NO. 1

PERCENTAGE OVERTIME CURVE APPLICABLE TO 1985
AND PROPOSED OVERTIME CURVE TARGET 1986

MANAGEMENT BY OBJECTIVES FOR THE YEAR 1986

MAINTENANCE DEPARTMENT

YARDSTICK FOR OBJECTIVE NO. 1

LABOUR OPTIMIZATION

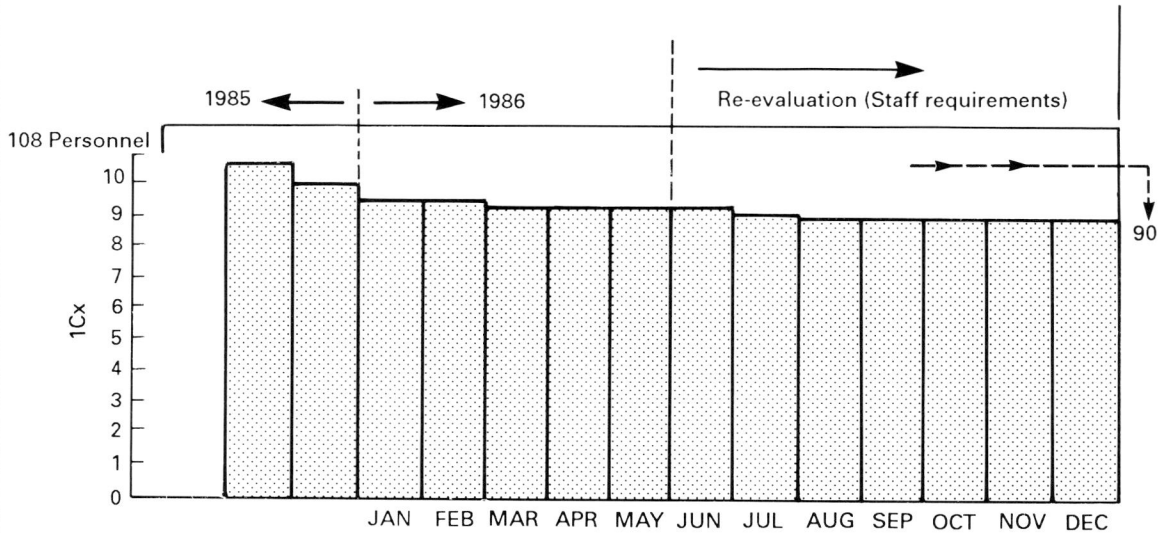

TOTAL 1986 MANPOWER

► 610000   8,571,755 DIRECT

► 620000   19,543,667 SUPPORT

TOTAL = 21,115,422

MAINTENANCE LABOUR REDUCTION ON THE CASCADE PRINCIPLE (1986) 9

TOTAL SAVED (APPROX.) 1,737,422.00

MANAGEMENT BY OBJECTIVES FOR THE YEAR

MAINTENANCE DEPARTMENT

YARDSTICKS FOR OBJECTIVE NO. 1

1986

MONTH	ACT. OT AVAIL. MH	APPROX. OT. (%)	1986 BUDGET (DH)	ACTUAL COST (DH)	SAVINGS (DHS)	REMARKS
JANUARY	$\frac{1435}{13579}$	10.6	57236	31000	26000	REVISED MAINTENANCE OUTAGE PROGRAMME 30 DAYS
FEBRUARY	$\frac{738}{13385}$	5.5	19832	16000	3800	GAS TURBINE OUTAGE PROGRAMME COMMENCE
MARCH	$\frac{433}{11340}$	4.0	8395	8395	–0–	COMPLETION OF GAS TURBINE MAINTENANCE OVERHAUL
APRIL	$\frac{200}{10986}$	2.0	10224	4000	6000	PREVENTIVE MAINTENANCE
MAY	$\frac{220}{11075}$	2.0	10224	4700	3500	PREVENTIVE MAINTENANCE
JUNE	$\frac{180}{8627}$	2.0	10224	3900	6300	HOLIDAY PERIOD
JULY	$\frac{173}{10751}$	1.61	8395	3800	4600	HOLIDAY PERIOD
AUGUST	$\frac{145}{7237}$	2.0	8395	3200	5200	HOLIDAY PERIOD
SEPTEMBER	$\frac{200}{10135}$	2.0	8395	4300	4000	
OCTOBER	$\frac{225}{11234}$	2.0	8395	4900	3500	
NOVEMBER	$\frac{277}{13852}$	2.0	57236	6000	51000	MAJOR OUTAGE PROGRAMME UNIT 2
DECEMBER	$\frac{226}{11312}$	2.0	19832	4900	15000	
TOTAL	4452		226783	95095	128900	

(1) TARGET AVERAGE ANNUAL OVERTIME FACTOR 3.142%.
(2) TARGET TOTAL OVERTIME HOURS 4,452.
(3) TARGET ACTUAL COST 95,095.

MANAGEMENT BY OBJECTIVES FOR THE YEAR 1986
MAINTENANCE DEPARTMENT
YARDSTICK FOR OBJECTIVE NO. 1

1985

MONTH	ACTUAL OT AVAIL M.H.	OVERTIME %	COST DHS	REMARKS
JANUARY	$\frac{1595}{13675}$	11.7	27115	VENDOR & WARRANTY PERIOD
FEBRUARY	$\frac{849}{13385}$	6.3	14433	OUTAGE
MARCH	$\frac{404}{11340}$	3.6	6868	
APRIL	$\frac{560}{10968}$	5.1	9520	
MAY	$\frac{391}{11075}$	3.5	6647	
JUNE	$\frac{353}{8627}$	4.1	6001	HOLIDAY PERIOD
JULY	$\frac{173}{10751}$	1.61	2941	HOLIDAY PERIOD
AUGUST	$\frac{349}{7237}$	4.8	5933	HOLIDAY PERIOD
SEPTEMBER	$\frac{543}{10135}$	5.4	9231	
OCTOBER	$\frac{906}{11234}$	8.1	15402	PREPARATION OF OUTAGE PROGRAMME
NOVEMBER	$\frac{1812}{13852}$	13.1	30804	OUTAGE PROGRAMME IN PROGRESS
DECEMBER	$\frac{707}{11312}$	6.3	12019	OUTAGE PROGRAMME NEAR COMPLETION
TOTAL			146914	

(1) AVERAGE ANNUAL OVERTIME FACTOR 6.134%.
(2) TOTAL OVERTIME HOURS 8642.
(3) TOTAL COST 146,914.

MANAGEMENT BY OBJECTIVES FOR THE YEAR

MAINTENANCE DEPARTMENT

OBJECTIVE NO.2

A 35 DAY MAJOR MAINTENANCE OUTAGE PROGRAMME
PER TRAIN WITH ZERO RATED OVERTIME.

1985 (NOV) OUTAGE PERIOD 50 DAYS PER TRAIN.

1985 (NOV) OVERTIME 1812 HOURS COST 30,804

MANAGEMENT BY OBJECTIVES FOR THE YEAR

MAINTENANCE DEPARTMENT

YARDSTICK FOR OBJECTIVE NO.2

(1) MAJOR OUTAGE NOVEMBER 1985

    (A) OUTAGE DURATION 2018 HOURS.

    (B) TOTAL OVERTIME 1812 HOURS.

    (C) BOILER AVAILABILITY 92%.

       DESAL. AVAILABILITY 92%.

       TURBINE AVAILABILITY 93.43%.

    NO VARIABLES BEING CONSIDERED.

(2) MAJOR OUTAGE JANUARY 1986

    (A) OUTAGE DURATION 1560 HOURS.

    (B) TOTAL OVERTIME 1435 HOURS.

    (C) BOILER AVAILABILITY 93.43%.

       DESAL. AVAILABILITY 94.00%.

       TURBINE AVAILABILITY 94.00%.

    NO VARIABLES BEING CONSIDERED.

(3) SCHEDULED MAJOR OUTAGE OCTOBER 1986

    (A) TARGET OUTAGE DURATION 1700 HOURS.

    (B) TARGET TOTAL OVERTIME ZERO.

    (C) BOILER AVAILABILITY 93.00%.

       DESAL. AVAILABILITY 94.00%.

       TURBINE AVAILABILITY 94.00%.

<u>MANAGEMENT BY OBJECTIVES FOR THE YEAR</u>

<u>MAINTENANCE DEPARTMENT</u>

<u>ACTION PLAN FOR OBJECTIVE NO.2</u>

S.No	ACTION	RESPONSIBILITY	INFORMATION	TARGET /DATE	
(1)	REGULAR MAINTENANCE/ PLANNING MEETINGS TO DISCUSS COMMON OBJECT- IVES, PROGRESS, APPLI- CABLE TO THE MAJOR MAINTENANCE OUTAGE OBJECTIVES/PHILOSOPH- IES.	MAINTENANCE SECTION HEADS/AREA ENGINEERS/ MAINTENANCE SUPERINTENDENT	PLANT MANAGER		
(2)	INVESTIGATION/FEASIBI- LITY STUDY INTO MAJOR MAINTENANCE OUTAGE PROGRAMME BY JOB CONTENT/PRIORITY.	MAINTENANCE SECTION HEADS/AREA ENGINEERS/ MAINTENANCE SUPERINTENDENT	PLANT MANAGER	MARCH 1986	
(3)	TO ASCERTAIN MAINTEN- ANCE OUTAGE JOB CONT- ENT BY ACTUAL PLANNED MAINTENANCE OUTAGE WORK AND PREVENTIVE MAINTENANCE WORK.	MAINTENANCE SECTION HEADS/AREA ENGINEERS/ MAINTENANCE SUPERINTENDENT	PLANT MANAGER	APRIL 1986	
(4)	TO PREPARE A REVISED MAINTENANCE OUTAGE SCHEDULE WHICH WILL ꝺꞓ ACTUAL PLANT OUTAGE MAINTENANCE AND PREVEN- TIVE MAINTENANCE. THE LATTER WHICH CAN BE INSTIGATED/ACCOMPLISHED	MAINTENANCE SECTION HEADS/AREA ENGINEERS/ MAINTENANCE SUPERINTENDENT	PLANT MANAGER	MAY 1986	

MANAGEMENT BY OBJECTIVES FOR THE YEAR

MAINTENANCE DEPARTMENT

YARDSTICK FOR OBJECTIVE NO. 2

1 ST. STAGE

1st Maint Outage

November 1985

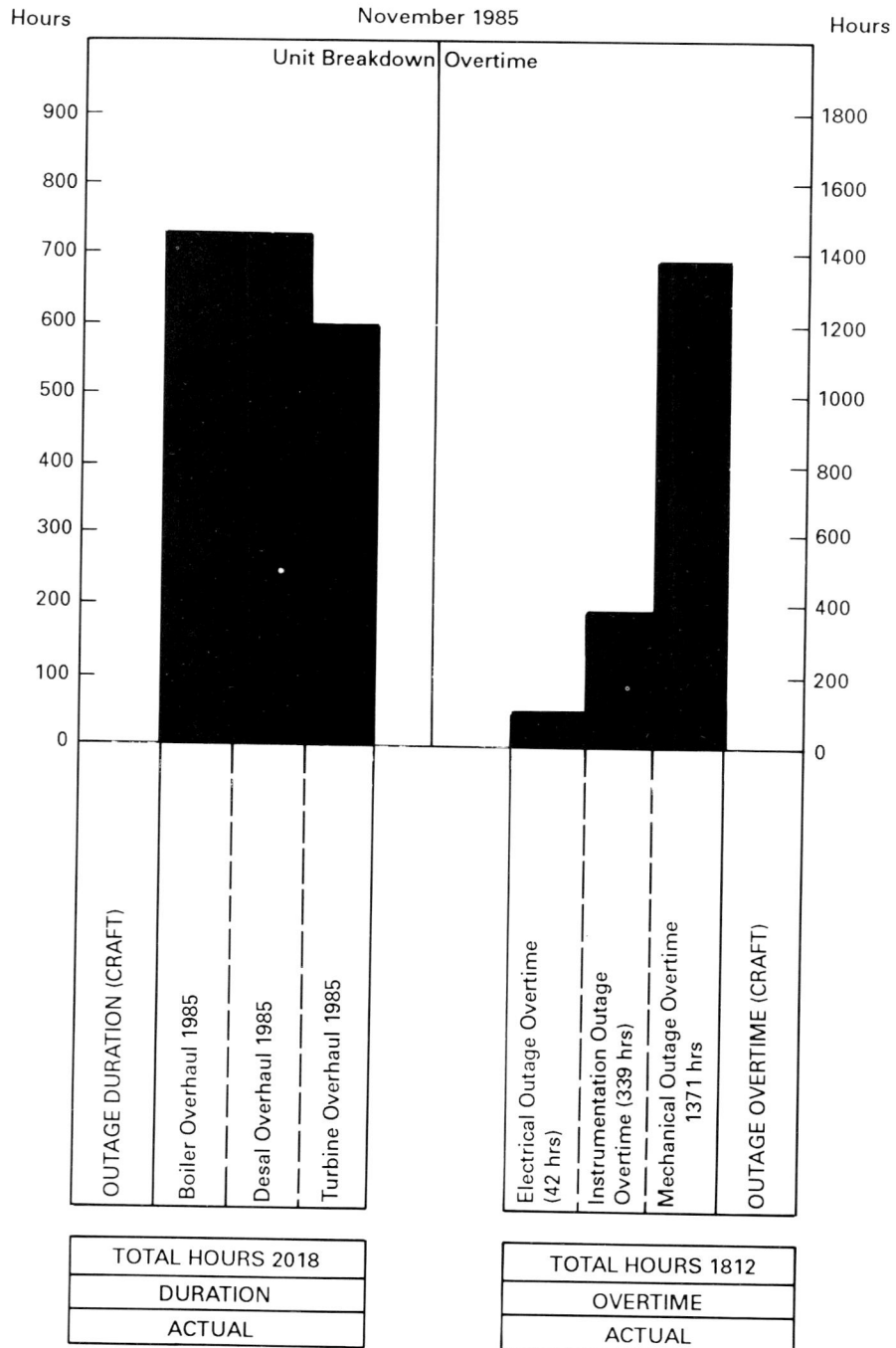

(1) MAINTENANCE OUTAGE PERIOD 2018 HOURS.
(2) OUTAGE OVERTIME 1812 HOURS.

MANAGEMENT BY OBJECTIVES FOR THE YEAR

MAINTENANCE DEPARTMENT

YARDSTICK FOR OBJECTIVE NO. 2

2ND STAGE
2nd Maint. Outage
January 1986

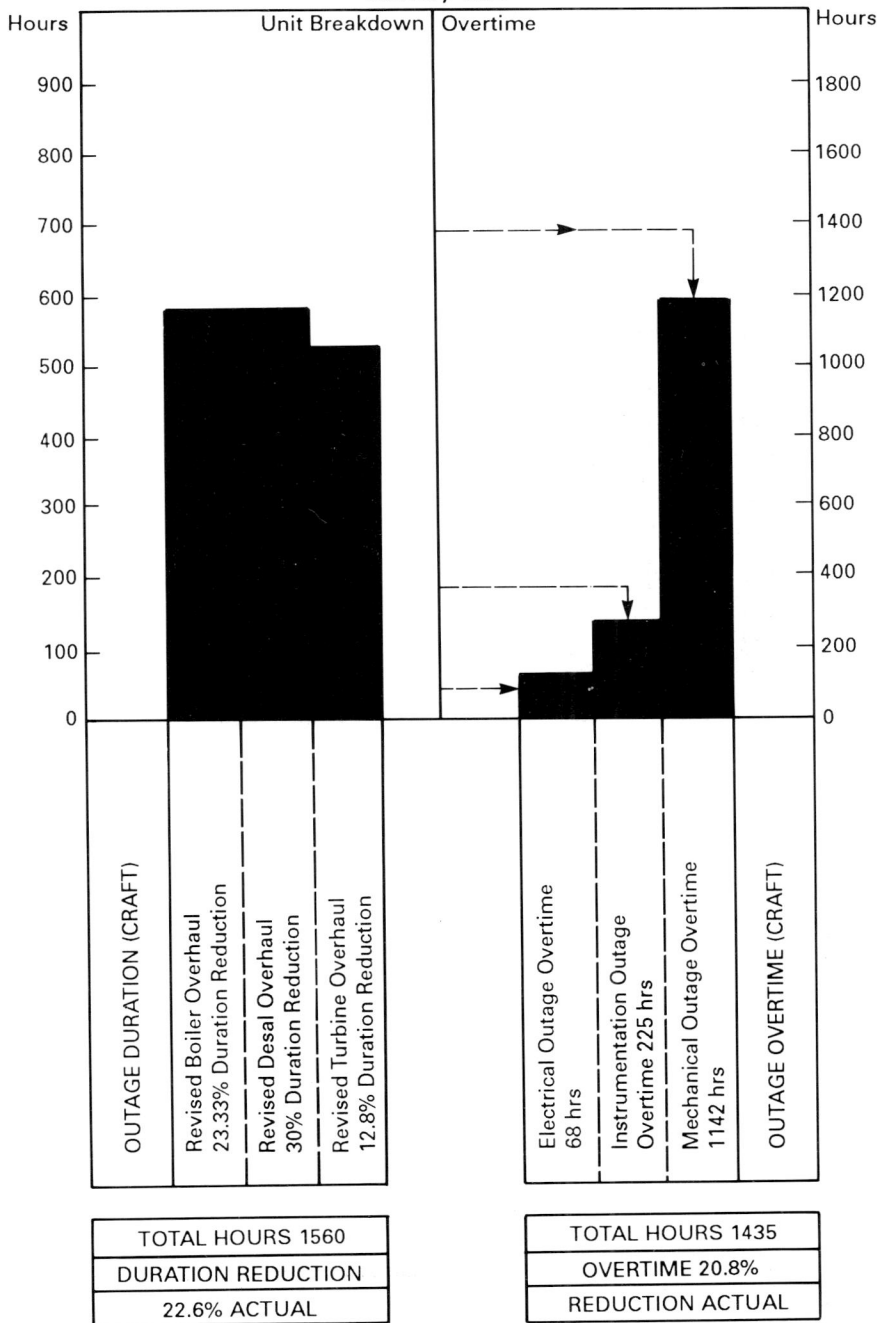

TOTAL HOURS 1560
DURATION REDUCTION
22.6% ACTUAL

TOTAL HOURS 1435
OVERTIME 20.8%
REDUCTION ACTUAL

(1) MAINTENANCE OUTAGE REDUCTION 22.6% = 1560 HOURS
(2) OUTAGE OVERTIME REDUCTION 20.8% = 1435 HOURS

MANAGEMENT BY OBJECTIVES FOR THE YEAR 1986

MAINTENANCE DEPARTMENT

YARDSTICK FOR OBJECTIVE NO. 2

3RD STAGE

3rd Maint Outage

November 1986

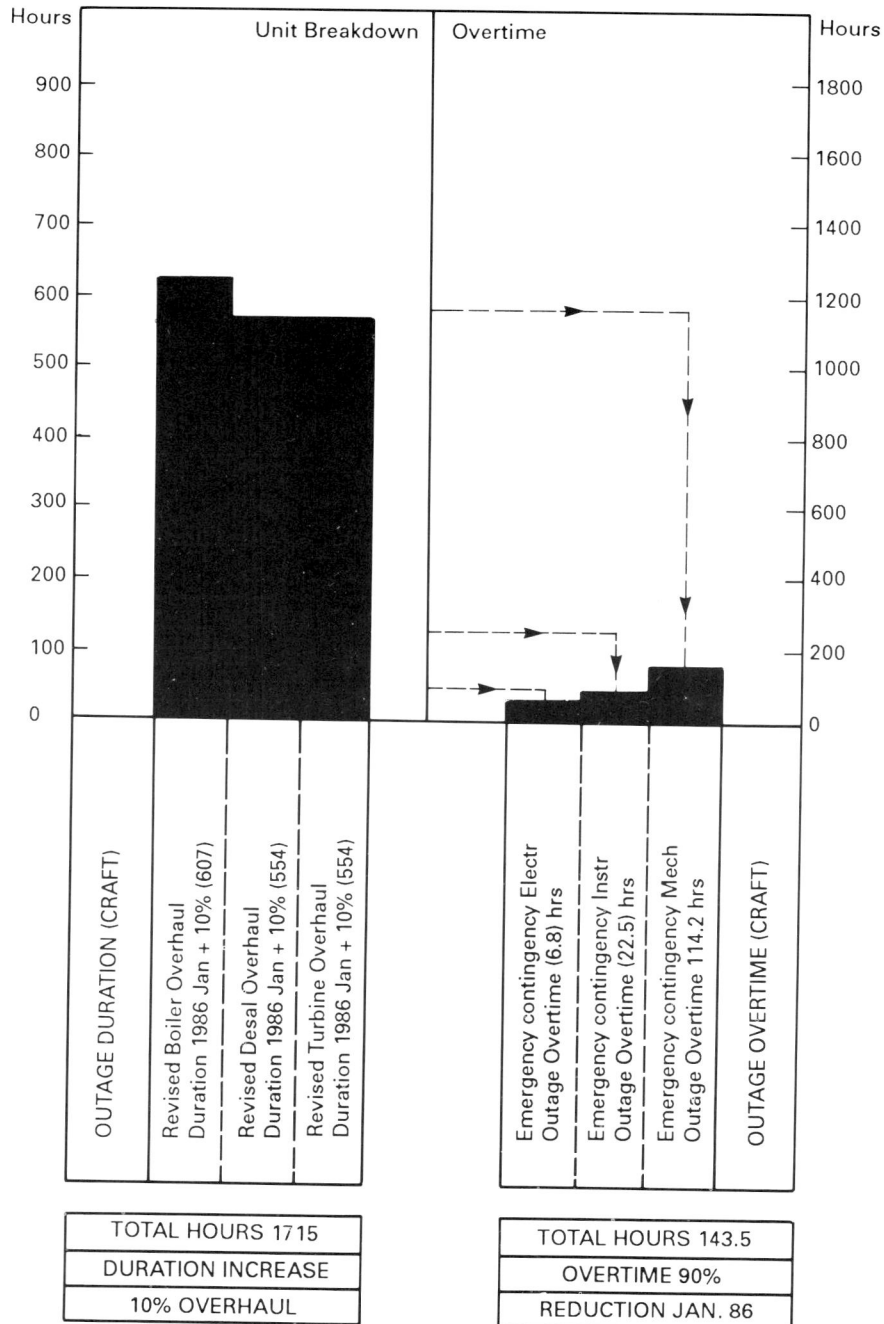

TOTAL HOURS 1715	TOTAL HOURS 143.5
DURATION INCREASE	OVERTIME 90%
10% OVERHAUL	REDUCTION JAN. 86

(1) MAINTENANCE OUTAGE DURATION INCREASED BY 10% (1716 HOURS).
(2) OUTAGE OVERTIME GROUND ZERO TARGET.

MANAGEMENT BY OBJECTIVES FOR THE YEAR

MAINTENANCE DEPARTMENT

OBJECTIVE NO.3

MAINTENANCE SPARE PARTS INVENTORY TO BE REDUCED
FROM 40,000,000 DIRHAMS TO 33,000,000 DIRHAMS
BY THE END OF 1986.

MONTHLY TARGET   583,333.00

MANAGEMENT BY OBJECTIVES FOR THE YEAR

MAINTENANCE DEPARTMENT

ACTION PLAN FOR OBJECTIVE NO.3

S.No.	A C T I O N	RESPONSIBILITY	INFORMATION	TARGET/ DATE
(1)	INVENTORY EVALUAT-ION CONTRACT ITEM	MATERIALS/MAINT. SECTION HEADS. MAINTENANCE SUPERINTENDENT	PLANT MANAGER	AUGUST 1986
(2)	INSURANCE ITEMS REVIEW	MATERIALS/MAINT. SECTION HEADS. MAINTENANCE SUPERINTENDENT	PLANT MANAGER	M A Y 1986
(3)	NON-MOVING ITEMS REVIEW	MATERIALS/MAINT. SECTION HEADS. MAINTENANCE SUPERINTENDENT	PLANT MANAGER	S E P T 1986
(4)	EXCESS/SURPLUS ITEMS REVIEW	MATERIALS/MAINT. SECTION HEADS. MAINTENANCE SUPERINTENDENT	PLANT MANAGER	D E C. 1986

MANAGEMENT BY OBJECTIVES FOR THE YEAR

MAINTENANCE DEPARTMENT

S.NO	ACTION	RESPONSIBILITY	INFORMATION	TARGET /DATE
	WHILE THE APPLICABLE UNIT REMAINS IN SERVICE, I.E. TURBINE CONDENSERS MAINTENANCE (SPLIT LEVEL) FEED PUMPS, ETC.			
(5)	INSTIGATION OF ALL PREVENTIVE MAINTENANCE PROCEEDING THE ACTUAL MAJOR MAINTENANCE OUTAGE WHILE PLANT IS IN OPERATION.	MAINTENANCE SECTION HEADS/AREA ENGINEERS/ MAINTENANCE SUPERINTENDENT.	PLANT MANAGER / OPERATIONS SUPERINTENDENT.	MAY 1986 ONWARDS

MANAGEMENT BY OBJECTIVES FOR THE YEAR

MAINTENANCE DEPARTMENT

YARDSTICK FOR OBJECTIVE NO. 3

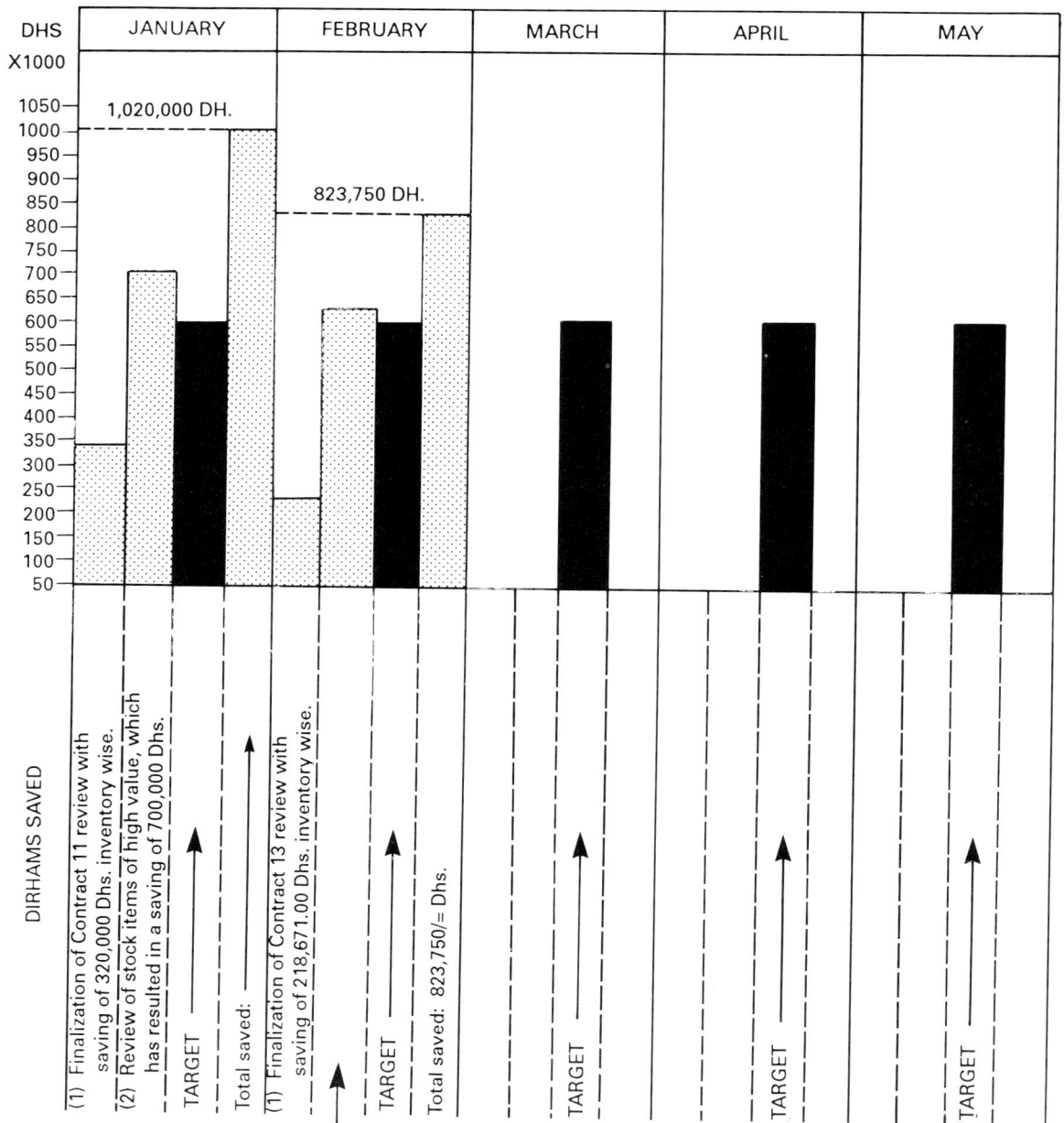

DHS X1000	JANUARY	FEBRUARY	MARCH	APRIL	MAY

1,020,000 DH.

823,750 DH.

1050—
1000—
950—
900—
850—
800—
750—
700—
650—
600—
550—
500—
450—
400—
350—
300—
250—
200—
150—
100—
50—

DIRHAMS SAVED

(1) Finalization of Contract 11 review with saving of 320,000 Dhs. inventory wise.

(2) Review of stock items of high value, which has resulted in a saving of 700,000 Dhs.

TARGET

Total saved:

(1) Finalization of Contract 13 review with saving of 218,671.00 Dhs. inventory wise.

TARGET

Total saved: 823,750/= Dhs.

TARGET

TARGET

TARGET

(2) Revision of Purchase Requisition and Stock Analysis sheet previously issued has resulted in a saving of 605,079.00 Dhs.

MONTHLY TARGET BASED OVER 12 MONTHS = 583,000.00 DHS. APPROX.

MANAGEMENT BY OBJECTIVES FOR THE YEAR

MAINTENANCE DEPARTMENT

ACTION PLAN FOR OBJECTIVE NO.2

		MAIN MAINTENANCE ACTIVITIES APPLICABLE TO THE PERIOD 1986/1987											

	MAINTENANCE ACTIVITY	JAN	FEB	MAR	APR	MAY	JUN	JUL	AUG	SEP	OCT	NOV	DEC
1	NO.1 BOILER MAINTENANCE/OVERHAUL	■											
2	NO.1 STEAM TURBINE MAINTENANCE OVERHAUL												
3	NO.1 DESAL. MAINTENANCE OVERHAUL	■											
4	G.E. GAS TURBINE MAINTENANCE OVERHAUL TP01		■										
5	FERTIL PUMPS/PLANT MAINTENANCE OVERHAUL				■								
6													
7													
8													
9													
10													
11													
12													
13													
14	PREVENTIVE MAINTENANCE NO.2 BOILER/TURBINE								■				
15	PREVENTIVE MAINTENANCE NO.1 BOILER/TURBINE								■				
16	NO.2 BOILER MAINTENANCE/OVERHAUL										■		
17	NO.2 STEAM TURBINE MAINTENANCE OVERHAUL											■	
18	NO.1 BOILER MAINTENANCE/OVERHAUL												
19	NO.1 STEAM TURBINE MAINTENANCE OVERHAUL												

1987

	MAINTENANCE ACTIVITY	JAN	FEB	MAR	APR	MAY	JUN	JUL	AUG	SEP	OCT	NOV	DEC
1	NO.1 DESAL MAINTENANCE/OVERHAUL	■											
2	NO.2 DESAL MAINTENANCE/OVERHAUL		■										
3													
4													
5	FERTIL PUMPS PLANT MAINTENANCE/OVERHAUL				■								
6													
7													
8													
9													
10													
11													
12													
13													
14													
15													
16	PREVENTIVE MAINTENANCE NO.2 BOILER/TURBINE								■				
17	PREVENTIVE MAINTENANCE NO.1 BOILER/TURBINE								■				
18	NO.2 BOILER MAINTENANCE/OVERHAUL										■		
19	NO.2 STEAM TURBINE											■	
20													
21													

(1)   MORE EMPHASIS/IMPORTANCE TO BE PLACED ON NON SHUT-DOWN MAINTENANCE, I.E. PREVENTIVE MAINTENANCE.

162

MANAGEMENT BY OBJECTIVES FOR THE YEAR

MAINTENANCE DEPARTMENT

OBJECTIVE NO.4

A REDUCTION OF 20% OF DEFECT ACCUMULATION BY
SEPTEMBER, 1986. THIS IS TO BE ACHIEVED BY
THE FULL IMPLEMENTATION OF THE PLANNED/PREVENT-
IVE MAINTENANCE SYSTEM.

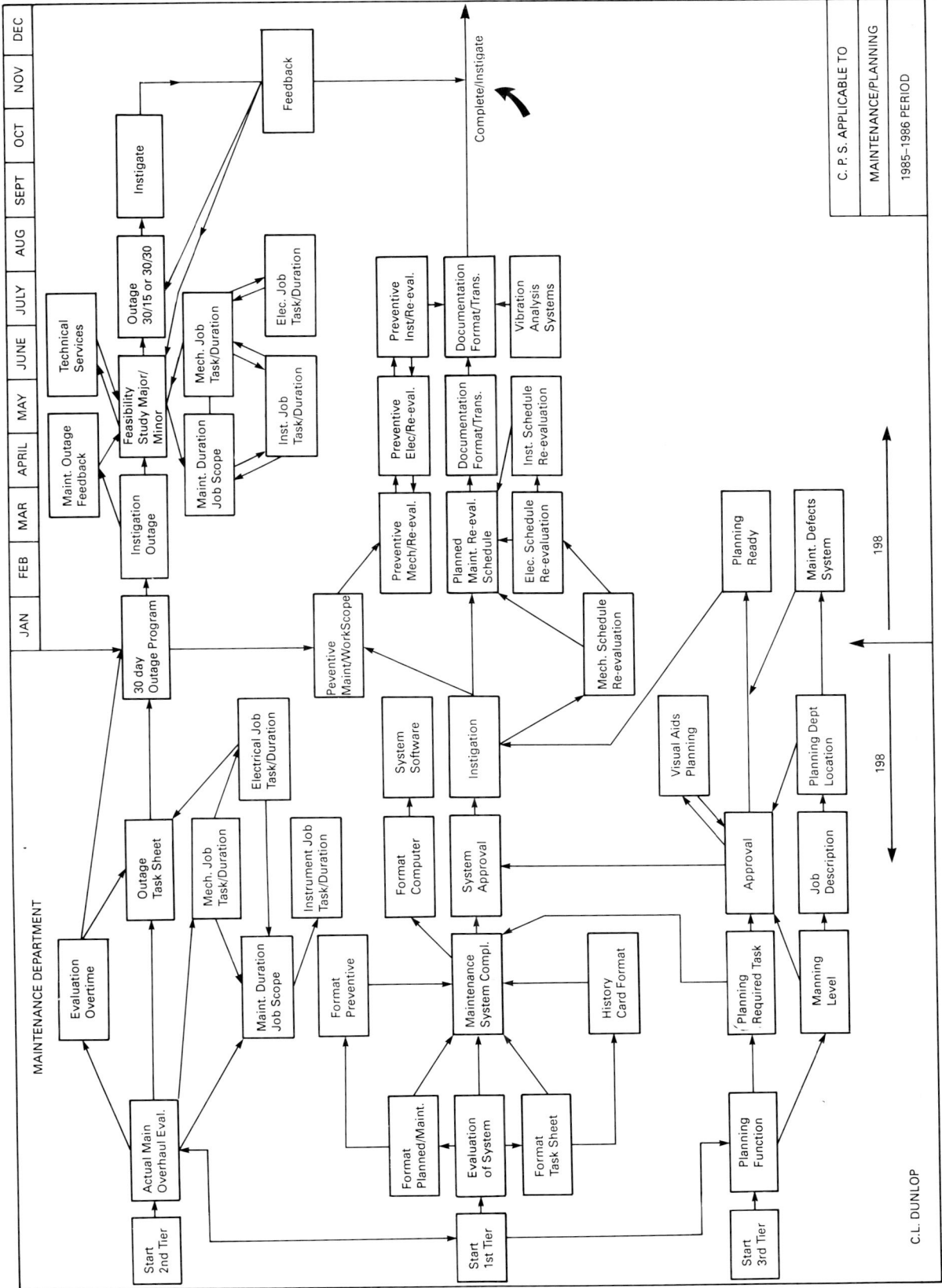

MAINTENANCE DEPARTMENT

JAN	FEB	MAR	APRIL	MAY	JUNE	JULY	AUG	SEPT	OCT	NOV	DEC

Feedback

Instigate

Technical Services

Outage 30/15 or 30/30

Maint. Outage Feedback

Feasibility Study Major/Minor

Instigation Outage

Maint. Duration Job Scope

Mech. Job Task/Duration

Elec. Job Task/Duration

Inst. Job Task/Duration

Complete/Instigate

30 day Outage Program

Peventive Maint/WorkScope

Preventive Mech/Re-eval.

Preventive Elec/Re-eval.

Preventive Inst/Re-eval.

Planned Maint. Re-eval. Schedule

Documentation Format/Trans.

Documentation Format/Trans.

Vibration Analysis Systems

Inst. Schedule Re-evaluation

Elec. Schedule Re-evaluation

Mech. Schedule Re-evaluation

System Software

Instigation

Planning Ready

Maint. Defects System

Evaluation Overtime

Outage Task Sheet

Mech. Job Task/Duration

Electrical Job Task/Duration

Maint. Duration Job Scope

Instrument Job Task/Duration

Format Preventive

Format Computer

System Approval

Visual Aids Planning

Approval

Job Description

Planning Dept Location

Actual Main Overhaul Eval.

Maintenance System Compl.

History Card Format

Planning Required Task

Manning Level

Format Planned/Maint.

Evaluation of System

Format Task Sheet

Planning Function

Start 2nd Tier

Start 1st Tier

Start 3rd Tier

198

198

C.L. DUNLOP

C. P. S. APPLICABLE TO

MAINTENANCE/PLANNING

1985–1986 PERIOD

MANAGEMENT BY OBJECTIVES FOR THE YEAR

MAINTENANCE DEPARTMENT

ACTION PLAN FOR OBJECTIVE NO.4

S.No.	ACTION	RESPONSIBILITY	INFORMATION	TARGET /DATE
(1)	THE RE-ORGANISATION OF THE PRESENT PLANNING SECTION WITH ADDITIONAL STAFF REPORTING STRUCTURE.	PLANNING SECTION HEAD/MAINTENANCE SUPERINTENDENT	PLANT MANAGER	APRIL 1986
(2)	THE INTRODUCTION OF REQUIRED PLANNED/ PREVENTIVE MAINT. DOCUMENTATION.	MAINTENANCE SUPERINTENDENT	PLANT MANAGER	FEB. 1986
(3)	THE RELOCATION OF ALL PLANNED/PREVE-NTIVE MAINTENANCE SCHEDULES/DOCUMENT-ATION FROM THE MECHANICAL,ELECTRI-CAL AND INSTRUMENT DEPARTMENTS TO THE CENTRAL PLANNING DEPARTMENT. ONE MAN FROM EACH DEP-ARTMENT TO BE PLANNING LIAISON CONTACT.	MAINTENANCE SECTION HEADS/ MAINTENANCE SUPERINTENDENT	PLANT MANAGER	JUNE 1986

MANAGEMENT BY OBJECTIVES FOR THE YEAR

MAINTENANCE DEPARTMENT

ACTION PLAN FOR OBJECTIVE NO.4

S.No	ACTION	RESPONSIBILITY	INFORMATION	TARGET /DATE
(4)	THE EVALUATION OF THE RELOCATED PLANNED/PREVENTIVE MAINTENANCE SCHED-ULES/DOCUMENTATION FROM CRAFT SECTI-ONS, I.E. OVER MAINTAINING.	MAINTENANCE SECTION HEADS/ MAINTENANCE SUPERINTENDENT	PLANT MANAGER	JUNE 1986
(5)	THE FORMATION/IMP-LEMENTATION OF A VIBRATION ANALYSIS SYSTEM: (A) IDENTIFY FIELD APPLICATIONS. (B) PREPARE REQUI-RED DOCUMENT-ATION. (C) TRAIN MAINT. PERSONNEL.	MECHANICAL SECTION HEAD/ MAINTENANCE SUPERINTENDENT	PLANT MANAGER	JULY 1986
(6)	THE INTRODUCTION OF A CENTRALIZED PLANNED/PREVENTIVE MAINTENANCE SYSTEM WITH OBJECTIVES AND GUIDELINES APPLICABLE TO ALL MAINTENANCE FUNCT-ION ACTIVITIES.	MAINTENANCE SUPERINTENDENT	PLANT MANAGER	JULY 1986

166

MANAGEMENT BY OBJECTIVES FOR THE YEAR

MAINTENANCE DEPARTMENT

YARDSTICK FOR OBJECTIVE NO. 4

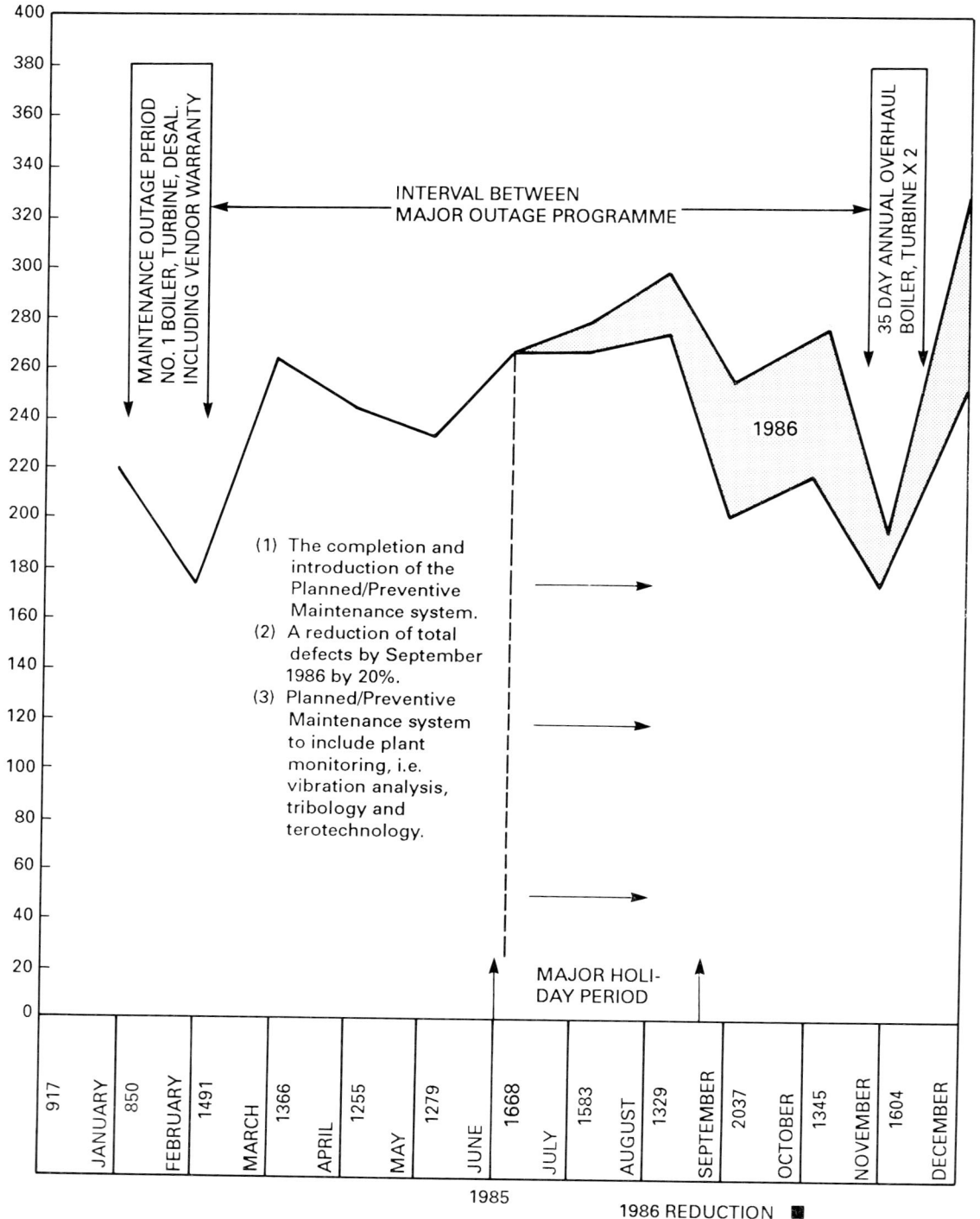

DEFECTS

400
380
360
340
320
300
280
260
240
220
200
180
160
140
120
100
80
60
40
20
0

MAINTENANCE OUTAGE PERIOD
NO. 1 BOILER, TURBINE, DESAL.
INCLUDING VENDOR WARRANTY

INTERVAL BETWEEN
MAJOR OUTAGE PROGRAMME

35 DAY ANNUAL OVERHAUL
BOILER, TURBINE X 2

1986

(1) The completion and
introduction of the
Planned/Preventive
Maintenance system.
(2) A reduction of total
defects by September
1986 by 20%.
(3) Planned/Preventive
Maintenance system
to include plant
monitoring, i.e.
vibration analysis,
tribology and
terotechnology.

MAJOR HOLI-
DAY PERIOD

| 917 | 850 | 1491 | 1366 | 1255 | 1279 | 1668 | 1583 | 1329 | 2037 | 1345 | 1604 | |
| JANUARY | FEBRUARY | MARCH | APRIL | MAY | JUNE | JULY | AUGUST | SEPTEMBER | OCTOBER | NOVEMBER | DECEMBER | |

1985

1986 REDUCTION ▪

MANAGEMENT BY OBJECTIVES FOR THE YEAR

MAINTENANCE DEPARTMENT

YARDSTICK FOR OBJECTIVE NO.4

MONTH	DEFECT CARDS NUMBERS MANHOURS				
	PRIORITY (A)	PRIORITY (B)	TOTAL NUMBERS	PRIORITY (A) MANHOURS	PRIORITY (B) MANHOURS
JANUARY	59	160	219	314	603
FEBRUARY	63	110	173	334	516
MARCH	70	191	261	644	847
APRIL	80	163	243	597	769
MAY	104	128	232	667	588
JUNE	70	193	263	440	839
JULY	110	167	1986 1985 263/277	790	878
AUGUST	77	213	275/290	406	1177
SEPTEMBER	105	148	233/253	717	612
OCTOBER	135	139	219/274	1268	769
NOVEMBER	86	108	155/194	709	636
DECEMBER	110	212	257/322	644	960
T O T A L	1069	1932	1986 1985 2793/3001	7530	9194

(1)   DEFECT CARD ACCUMULATION 1985 (3001)

(2)   DEFECT CARD REDUCTION APRIL 1986 (2793)

MAJOR OUTAGE MAINTENANCE

PROGRAMME OBJECTIVES/TARGET

NOTE

(1) The major maintenance outage programme no.1 was completed in November, 1985 and was monitored by the new Maintenance Superintendent for necessary modification.

(2) The major maintenance outage programme no.2 was completed in January, 1986 and met all objectives/targets as stipulated (A) Reduction in overtime, (B) Reduction in outage duration, (C) Optimized outage maintenance.

(3) The major maintenance outage programme no.3 is scheduled with the following objectives/targets and is based on two maintenance outages per year of 35 days duration each. There will be an actual increase of the outage duration as measured against the no.2 outage by 10% with a ground zero rated overtime factor. A emergency overtime contingency of 10% based on the no.2 outage overtime accumulation has been installed.

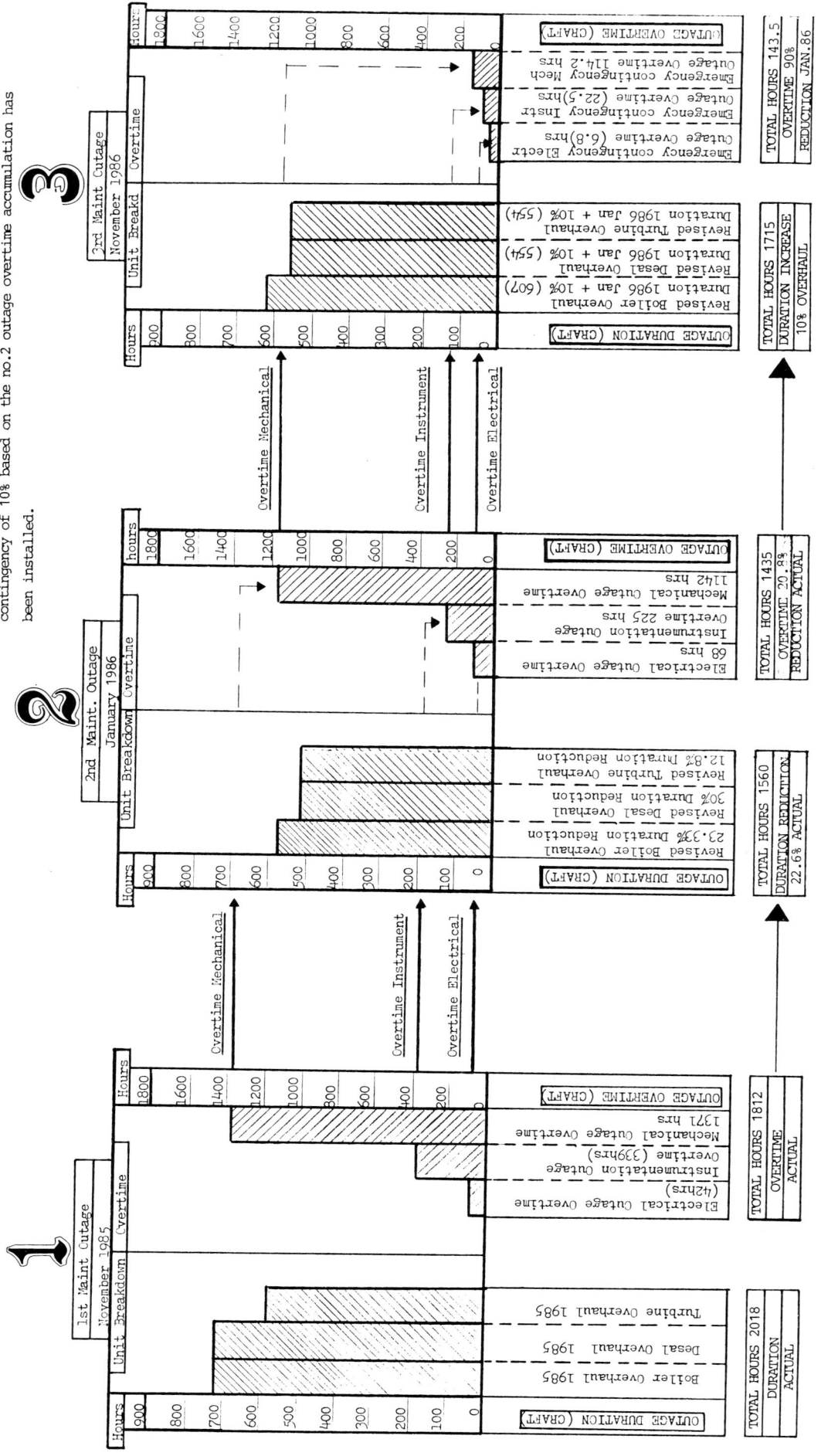

---

① 1st Maint Cutage November 1985 — Unit Breakdown / Overtime

OUTAGE DURATION (CRAFT)
- Boiler Overhaul 1985
- Desal Overhaul 1985
- Turbine Overhaul 1985

TOTAL HOURS 2018 — DURATION — ACTUAL

Overtime Mechanical / Overtime Instrument / Overtime Electrical

OUTAGE OVERTIME (CRAFT)
- Electrical Outage Overtime (42hrs)
- Instrumentation Outage Overtime (339hrs)
- Mechanical Outage Overtime 1371 hrs

TOTAL HOURS 1812 — OVERTIME — ACTUAL

---

② 2nd Maint. Outage January 1986 — Unit Breakdown / Overtime

OUTAGE DURATION (CRAFT)
- Revised Boiler Overhaul 23.3% Duration Reduction
- Revised Desal Overhaul 30% Duration Reduction
- Revised Turbine Overhaul 12.8% Duration Reduction

TOTAL HOURS 1560 — DURATION REDUCTION — 22.6% ACTUAL

Overtime Mechanical / Overtime Instrument / Overtime Electrical

OUTAGE OVERTIME (CRAFT)
- Electrical Outage Overtime 68 hrs
- Instrumentation Outage Overtime 225 hrs
- Mechanical Outage Overtime 1142 hrs

TOTAL HOURS 1435 — OVERTIME 20.8% — REDUCTION ACTUAL

---

③ 2nd Maint Cutage November 1986 — Unit Breakd / Overtime

OUTAGE DURATION (CRAFT)
- Revised Boiler Overhaul Duration 1986 Jan + 10% (607)
- Revised Desal Overhaul Duration 1986 Jan + 10% (554)
- Revised Turbine Overhaul Duration 1986 Jan + 10% (554)

TOTAL HOURS 1715 — DURATION INCREASE — 10% OVERHAUL

OUTAGE OVERTIME (CRAFT)
- Emergency contingency Electr Outage Overtime (6.8) hrs
- Emergency contingency Instr Outage Overtime (22.5) hrs
- Emergency contingency Mech Outage Overtime 114.2 hrs

TOTAL HOURS 143.5 — OVERTIME 90% — REDUCTION JAN.86

DATE:

WATER & POWER STATION
MAINTENANCE DEPARTMENT

REVISED & PROPOSED MAINTENANCE OUTAGE/OPERATIONAL
OVERTIME APPLICABLE TO 1986 PERIOD

PERCENTAGE OVERTIME CURVE APPLICABLE TO 1985
AND PROPOSED OVERTIME CURVE TARGET 1986

1986
TARGET/FORECAST

NOTE: HIDDEN SAVINGS NOT MENTIONED, I.E.
REDUCTION OF CONTRACT HELPERS BY
NUMBERS AND DURATION.

MONTH	ACT. OT AVAIL. MH	APPROX.OT (%)	1986 BUDGET (DH)	ACTUAL COST (DH)	SAVINGS (DHS)	R E M A R K S
JANUARY	1435 / 13579	10.6	57236	31000	26000	REVISED MAINTENANCE OUTAGE PROGRAMME 30 DAYS
FEBRUARY	738 / 13385	5.5	19832	16000	3800	GAS TURBINE OUTAGE PROGRAMME COMMENCE
MARCH	433 / 11340	4.0	8395	8395	-0-	COMPLETION OF GAS TURBINE MAINTENANCE OVERHAUL.
APRIL	200 / 10968	2.0	10224	4000	6000	PREVENTIVE MAINTENANCE
MAY	220 / 11075	2.0	10224	4700	3500	PREVENTIVE MAINTENANCE
JUNE	180 / 8627	2.0	10224	3900	6300	HOLIDAY PERIOD
JULY	173 / 10751	1.61	8395	3800	4600	HOLIDAY PERIOD
AUGUST	145 / 7237	2.0	8395	3200	5200	HOLIDAY PERIOD
SEPTEMBER	200 / 10135	2.0	8395	4300	4000	
OCTOBER	225 / 11234	2.0	8395	4900	3500	
NOVEMBER	277 / 13852	2.0	57236	6000	51000	MAJOR OUTAGE PROGRAMME UNIT 2.
DECEMBER	226 / 11312	2.0	19832	4900	15000	
T O T A L	4452		226783	95095	128900	

(1) The above mentioned composes of two maintenance outages per year 2 at 35 days
complete.

(2) As stated in proposed maintenance objectives/target approx. saving of approx.
40% for the 1986 fiscal year is visualised.

C.L. DUNLOP
MAINTENANCE SUPERINTENDENT

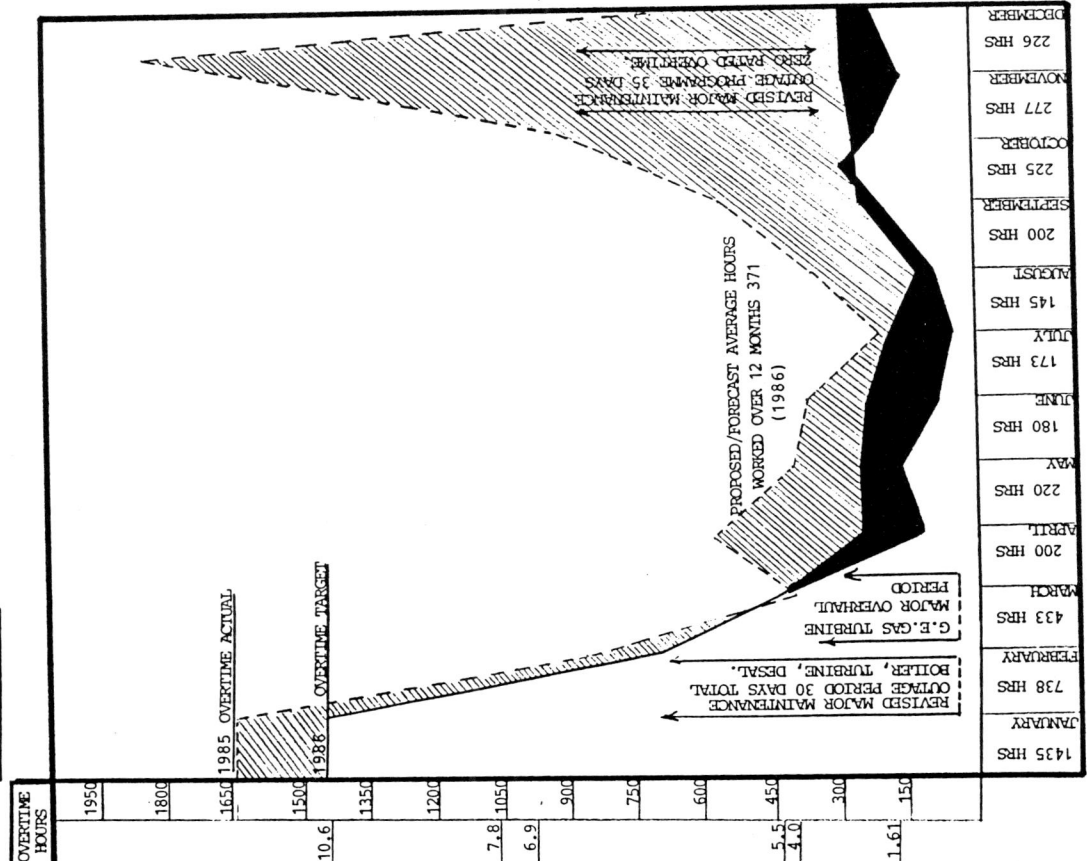

REVISED MAJOR MAINTENANCE
OUTAGE PERIOD 30 DAYS TOTAL.
BOILER, TURBINE, DESAL.

G.E.GAS TURBINE
MAJOR OVERHAUL
PERIOD

REVISED MAJOR MAINTENANCE
OUTAGE PROGRAMME 35 DAYS
ZERO RATED OVERTIME.

PROPOSED/FORECAST AVERAGE HOURS
WORKED OVER 12 MONTHS 371
(1986)

1985 OVERTIME ACTUAL
1986 OVERTIME TARGET

OVERTIME HOURS: 1950, 1800, 1650, 1500, 1350, 1200, 1050, 900, 750, 600, 450, 300, 150

| JANUARY 1435 HRS | FEBRUARY 738 HRS | MARCH 433 HRS | APRIL 200 HRS | MAY 220 HRS | JUNE 180 HRS | JULY 173 HRS | AUGUST 145 HRS | SEPTEMBER 200 HRS | OCTOBER 225 HRS | NOVEMBER 277 HRS | DECEMBER 226 HRS |

MAINTENANCE DEPARTMENT

DEFECT TARGET

DEFECT ACCUMULATION WITH APPLICABLE MANHOURS FOR REQUIRED RECTIFICATION

DEFECT ACCUMULATION REDUCTION

PRIORITY A & B

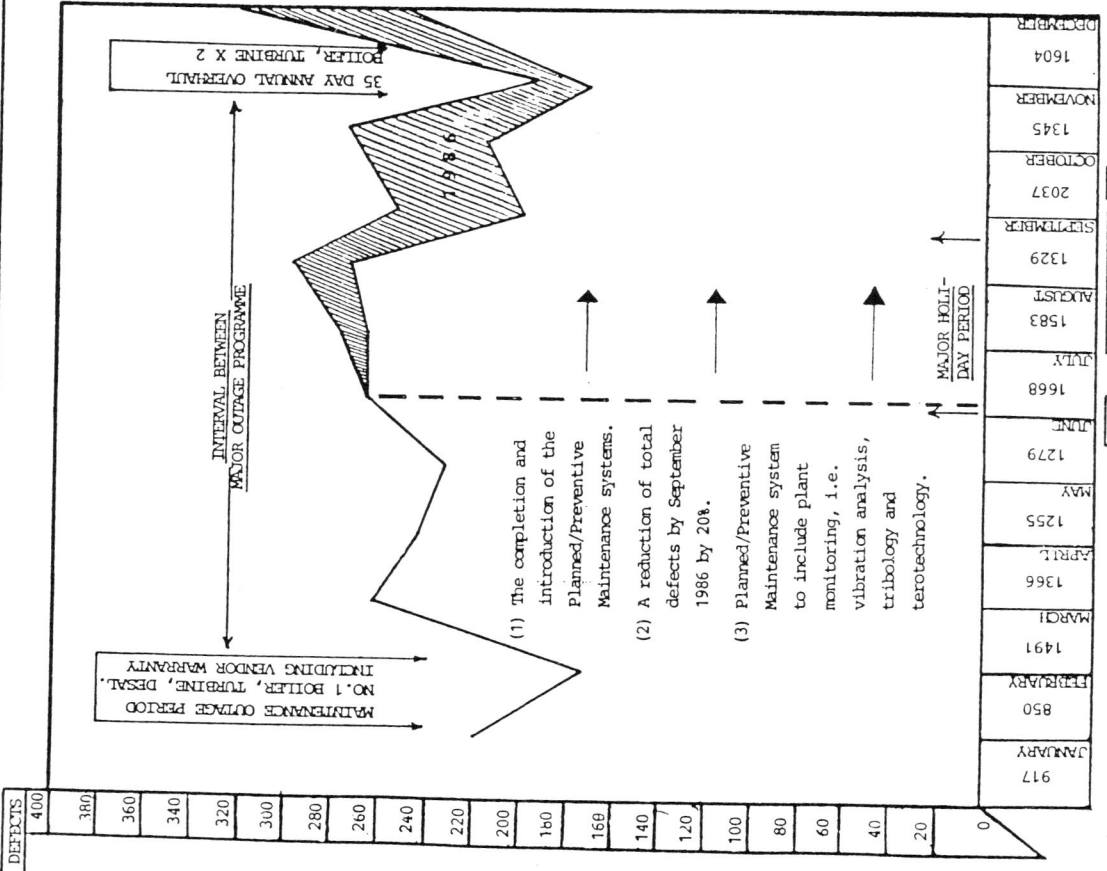

			DEFECT CARDS NUMBERS MANHOURS		
MONTH	PRIORITY (A)	PRIORITY (B)	TOTAL NUMBERS	PRIORITY(A) MANHOURS	PRIORITY(B) MANHOURS
JANUARY	59	160	219	314	603
FEBRUARY	63	110	173	334	516
MARCH	70	191	261	644	847
APRIL	80	163	243	597	769
MAY	104	128	232	667	588
JUNE	70	193	263	440	839
JULY	110	167	1986/1985 263/277	790	878
AUGUST	77	213	275/290	406	1177
SEPTEMBER	105	148	233/253	717	612
OCTOBER	135	139	219/274	1268	769
NOVEMBER	86	108	155/194	709	636
DECEMBER	110	212	1986/1985 257/322	644	960
TOTAL	1069	1932	1986/1985 2793/3001	7530	9194

NOTE: WPS has no work order/requests for service as applicable to other HP sites. All services to Operations are logged on a Defect Card, i.e. 'Chlorine cylinder change out', etc.

DEFECTS

INTERVAL BETWEEN MAJOR OUTAGE PROGRAMME

35 DAY ANNUAL OVERHAUL BOILER, TURBINE X 2

1986

MAINTENANCE OUTAGE PERIOD NO.1 BOILER, TURBINE, DESAL, INCLUDING VENDOR WARRANTY

MAJOR HOLIDAY PERIOD

1985   1986 REDUCTION

(1) The completion and introduction of the Planned/Preventive Maintenance systems.

(2) A reduction of total defects by September 1986 by 20%.

(3) Planned/Preventive Maintenance system to include plant monitoring, i.e. vibration analysis, tribology and terotechnology.

JANUARY	FEBRUARY	MARCH	APRIL	MAY	JUNE	JULY	AUGUST	SEPTEMBER	OCTOBER	NOVEMBER	DECEMBER
917	850	1491	1366	1255	1279	1668	1583	1329	2037	1345	1604

WATER & POWER STATION
MAINTENANCE DEPARTMENT
DEFECT TARGET 1986

13TH JANUARY, 1986
PRIORITY A & B

DEFECT ACCUMULATION WITH APPLICABLE MANHOURS FOR REQUIRED RECTIFICATION (1985)
1986 DEFECT ACCUMULATION REDUCTION

DEFECT CARDS NUMBERS MANHOURS

MONTH	PRIORITY (A)	PRIORITY (B)	TOTAL NUMBERS	PRIORITY(A) MANHOURS	PRIORITY(B) MANHOURS
JANUARY	59	160	219	314	603
FEBRUARY	63	110	173	334	516
MARCH	70	191	261	644	847
APRIL	80	163	243	597	769
MAY	104	128	232	667	588
JUNE	70	193	263	440	839
JULY	110	167	1986 1985 263/277	790	878
AUGUST	77	213	275/290	406	1177
SEPTEMBER	105	148	233/253	717	612
OCTOBER	135	139	219/274	1268	769
NOVEMBER	86	108	155/194	709	636
DECEMBER	110	212	1986 1985 257/322	644	960
TOTAL	1069	1932	1986 1985 2793/3001	7530	9194

NOTE: WPS has no work order/requests for service
as applicable to other HP sites. All
services to Operations are logged on a
Defect Card, i.e. 'Chlorine cylinder change
out', etc.

C.M. DUNLOP

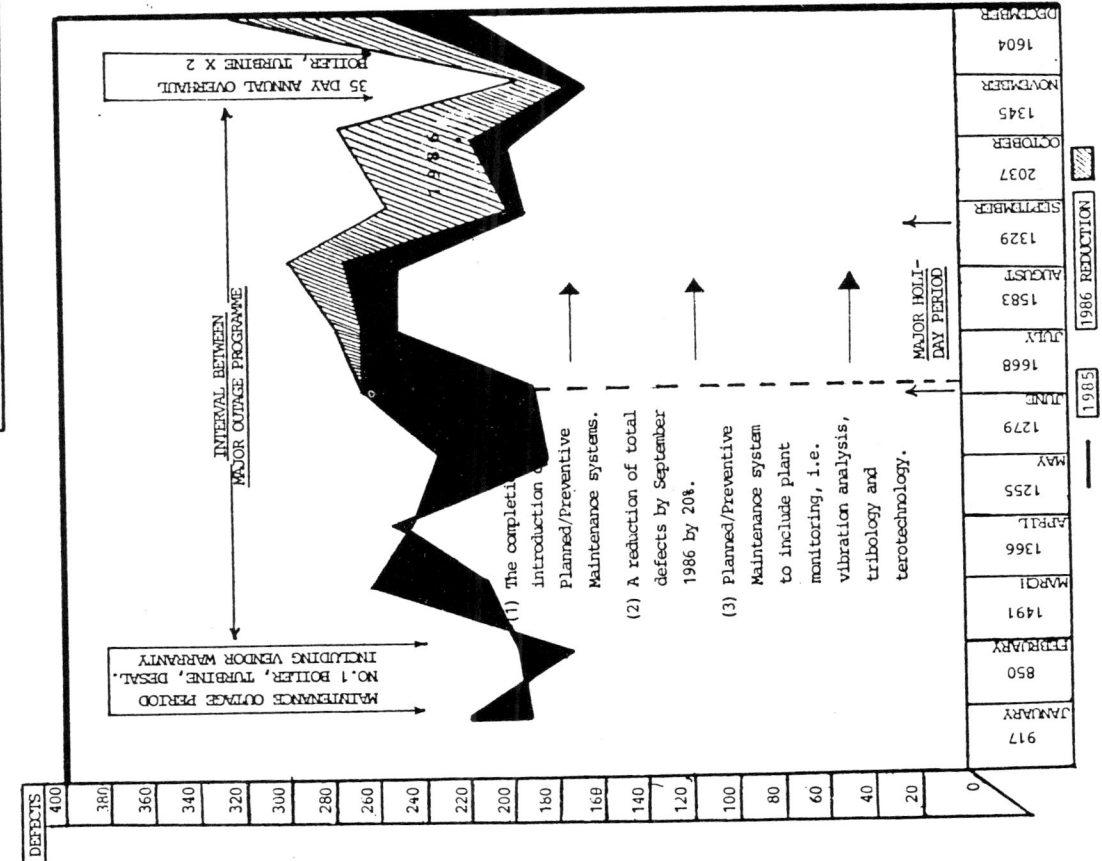

DEFECTS

400
380
360
340
320
300
280
260
240
220
200
180
160
140
120
100
80
60
40
20
0

JANUARY 917
FEBRUARY 850
MARCH 1491
APRIL 1366
MAY 1255
JUNE 1279
JULY 1668
AUGUST 1583
SEPTEMBER 1329
OCTOBER 2037
NOVEMBER 1345
DECEMBER 1604

1985
1986 REDUCTION

MAINTENANCE OUTAGE PERIOD
NO.1 BOILER, TURBINE, DESAL,
INCLUDING VENDOR WARRANTY

INTERVAL BETWEEN
MAJOR OUTAGE PROGRAMME

35 DAY ANNUAL OVERHAUL
BOILER, TURBINE X 2

1985

MAJOR HOLI-
DAY PERIOD

(1) The completion &
introduction of
Planned/Preventive
Maintenance systems.

(2) A reduction of total
defects by September
1986 by 20%.

(3) Planned/Preventive
Maintenance system
to include plant
monitoring, i.e.
vibration analysis,
tribology and
terotechnology.

EXAMPLE OF MAINTENANCE DEPARTMENT
FOR A MEDIUM/LARGE POWER STATION

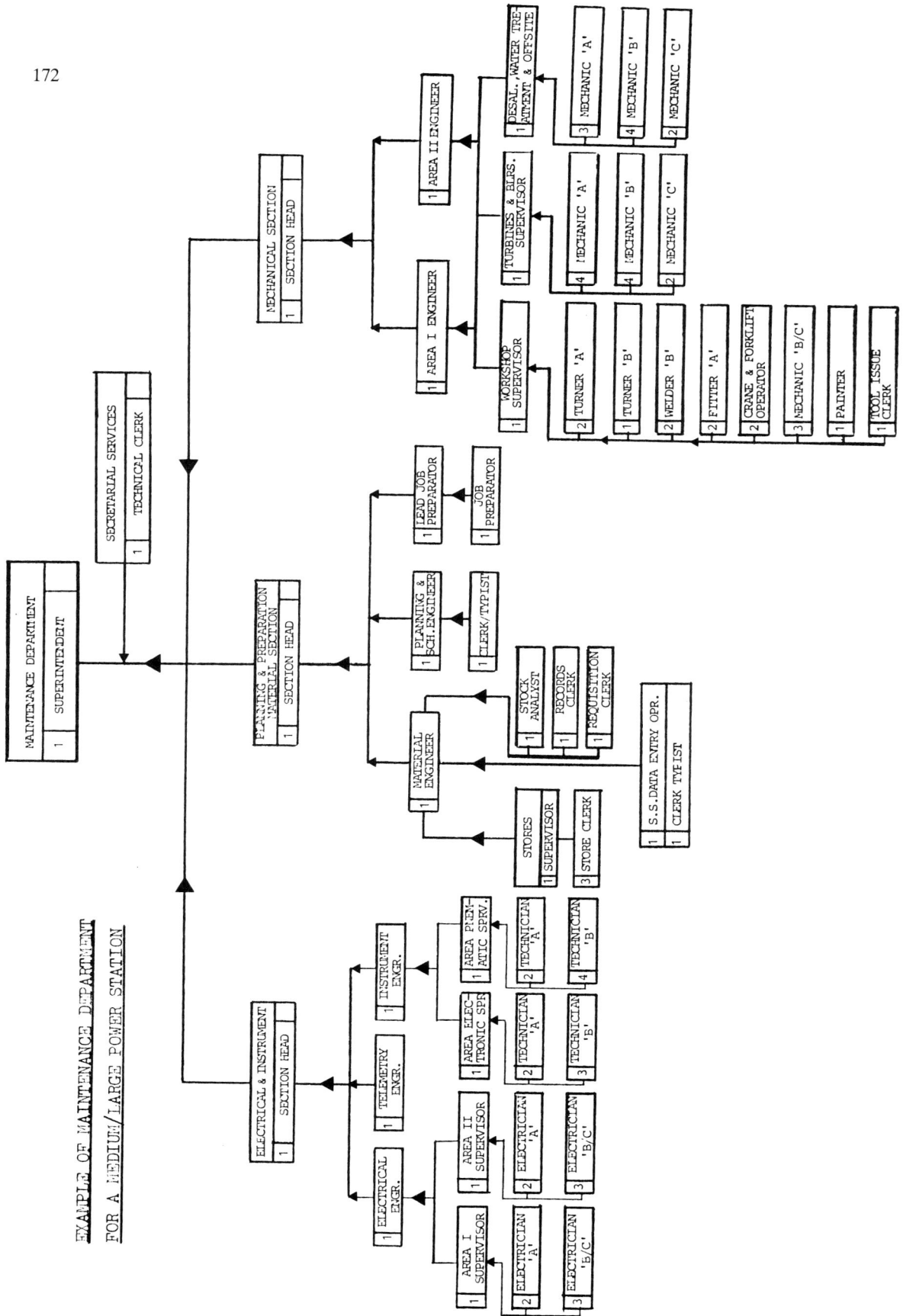

MAINTENANCE DEPARTMENT — SUPERINTENDENT — 1

SECRETARIAL SERVICES — TECHNICAL CLERK — 1

MECHANICAL SECTION — SECTION HEAD — 1

AREA II ENGINEER — 1

DESAL, WATER TREATMENT & OFFSITE — 1
- MECHANIC 'A' — 3
- MECHANIC 'B' — 4
- MECHANIC 'C' — 2

TURBINES & BLRS. SUPERVISOR — 1
- MECHANIC 'A' — 4
- MECHANIC 'B' — 4
- MECHANIC 'C' — 2

AREA I ENGINEER — 1

WORKSHOP SUPERVISOR — 1
- TURNER 'A' — 2
- TURNER 'B' — 1
- WELDER 'B' — 2
- FITTER 'A' — 2
- CRANE & FORKLIFT OPERATOR — 2
- MECHANIC 'B/C' — 3
- PAINTER — 1
- TOOL ISSUE CLERK — 1

PLANNING & PREPARATION / MATERIAL SECTION — SECTION HEAD — 1

PLANNING & SCH.ENGINEER — 1
- LEAD JOB PREPARATOR — 1
- JOB PREPARATOR — 1
- CLERK/TYPIST — 1

MATERIAL ENGINEER — 1
- STOCK ANALYST — 1
- RECORDS CLERK — 1
- REQUISITION CLERK — 1
- S.S.DATA ENTRY OPR. — 1
- CLERK TYPIST — 1

STORES SUPERVISOR — 1
- STORE CLERK — 3

ELECTRICAL & INSTRUMENT — SECTION HEAD — 1

INSTRUMENT ENGR. — 1

AREA PNEUMATIC SPRV. — 2
- TECHNICIAN 'A' — 2
- TECHNICIAN 'B' — 4

AREA ELECTRONIC SPRV. — 1
- TECHNICIAN 'A' — 2
- TECHNICIAN 'B' — 3

TELEMETRY ENGR. — 1

ELECTRICAL ENGR. — 1

AREA II SUPERVISOR — 1
- ELECTRICIAN 'A' — 2
- ELECTRICIAN 'B/C' — 3

AREA I SUPERVISOR — 1
- ELECTRICIAN 'A' — 2
- ELECTRICIAN 'B/C' — 3

CONCLUSION

The basis of Scheduled Maintenance philosophy is that
of a service rather than an organisation. By this, it
is meant that the Planned Maintenance will help Maintenance Engineers to establish and operate such a system
to provide a high plant availability factor.

To permit flexibility, the main object will be to keep
the system as simple as possible.

Although my system appears to be complicated when described in length, the inter-relationship between the
necessary documentation is basically simple.